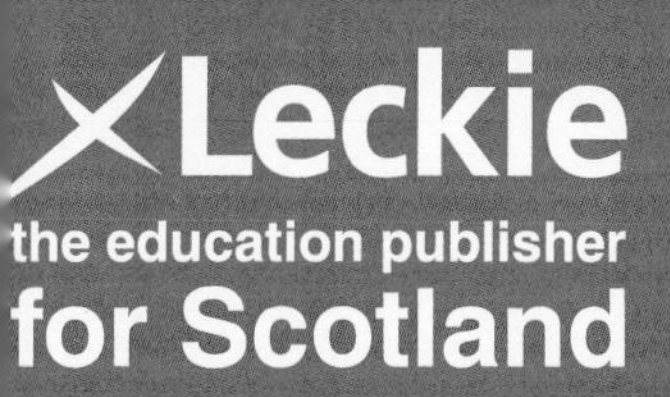

National 5 PHYSICS

For SQA 2019 and beyond

Revision + Practice
2 Books in 1

001/01102020

10 9 8 7 6 5 4 3 2 1

ISBN 9780008435363

Published by
Leckie
An imprint of HarperCollinsPublishers
Westerhill Road, Bishopbriggs, Glasgow, G64 2QT
T: 0844 576 8126
leckiescotland@harpercollins.co.uk www.leckiescotland.co.uk

Publisher: Sarah Mitchell
Project Manager: Harley Griffiths and Lauren Murray

Special thanks to
QBS (layout and illustration)

Printed in Italy by Grafica Veneta S.p.A

A CIP Catalogue record for this book is available from the British Library.

To access the ebook version of this Revision Guide visit
www.collins.co.uk/ebooks
and follow the step-by-step instructions.

Introduction 7

National 5 physics and equations 8

Specified data 13

Part 1: Revision guide

Area 1: Dynamics

Scalars and vectors 16

Measuring displacement and velocity 18

Resultant vectors 20

Velocity-time graphs 22

Velocity and displacement 24

Acceleration 26

Acceleration graphs 28

Newton's 1st law of motion 30

Newton's 2nd law of motion 32

Forces in action 34

Newton's 3rd law of motion 36

Newton's laws in action 38

Matter and energy 40

Work done 42

Potential and kinetic energy 44

Energy conservation 46

Mass and weight 48

Projectile motion 50

Contents

Area 2: Space

Satellites	52
Space exploration	54
Cosmology	56
Spectra	58

Area 3: Electricity

Electrical charge and current	60
d.c. and a.c.	62
Electric fields	64
Voltage	66
Resistance and Ohm's Law	68
Components and symbols	70
Electronic circuits	72
Series circuits	74
Parallel circuits	76
Energy and power	78
Power and fuses	80

Area 4: Properties of matter

Heat	82
Heat capacity	84
Latent heat	86
Pressure	88
Gas: pressure - temperature	90
Gas: volume - temperature	92
Final gas laws	94

Area 5: Waves

Waves 96

Wave equations 1 98

Wave equations 2 100

Diffraction 102

Electromagnetic spectrum: facts 104

Electromagnetic spectrum: data 106

Reflection and Refraction 108

Angles of refraction 110

Area 6: Radiation

Nuclear nature 112

Ionisation 114

Nuclear activity 116

Nuclear applications 118

Nuclear relationships 120

Half-life 122

Nuclear fission 126

Nuclear fusion 128

Scientific values and exam guide

Units, prefixes and scientific notation 130

Exam question guide 132

Contents

Part 2: Practice test papers

Introduction	136
Practice paper A	140
Practice paper B	173

ANSWERS Check your answers to the practice test papers online: www.leckieandleckie.co.uk

Introduction

Complete Revision and Practice

This Complete **two-in-one Revision and Practice** book is designed to support you as students of National 5 Physics. It can be used either in the classroom, for regular study and homework, or for exam revision. By combining **a revision guide and two full sets of practice exam papers**, this book includes everything you need to be fully familiar with the National 5 Physics exam. As well as including ALL the core course content with practice opportunities, there is comprehensive assignment and exam preparation advice with both revision question and practice test paper answers provided online at www.leckieandleckie.co.uk.

Introducing physics

The universe consists of matter, energy, space and time. Studying physics deepens our understanding of these and the interaction between them. We learn to ask questions, to observe and experiment. As well as learning more of the impact of physics on everyday life we develop scientific skills, thinking skills, technological skills, problem-solving skills, and independent working skills as well as being able to make scientifically informed choices.

- Studying physics provides us with a deep understanding of our world and its place in the Universe.
- An understanding of physics is fundamental to a deeper understanding of all science.
- Physics studies the energy sources we need for everyday life through to the latest developments in the exploration of space and, in between, all the many applications that have been developed as a result of the discoveries of the laws of physics.
- Physics has been developed as a result of practical experimentation and theoretical thinking, and you will develop these skills as you do this physics course.
- In scale, physics ranges from the study of the smallest parts of the atom through to the size of the Universe itself.
- Modern technology exists and develops as a result of our understanding of physics.

Traffic lights

You will find traffic lights throughout this book to help you assess how well you understand each section.

GOT IT? ☐ ☐ ☐

Study hard and success will follow.

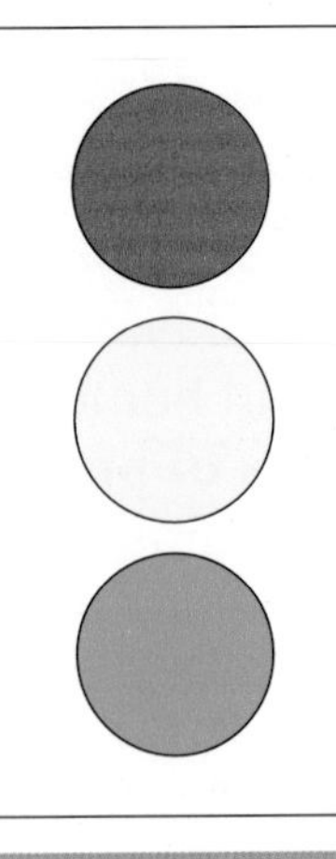

I don't get it! I need some help understanding.

I think I understand but I need a little support.

I understand and can try this on my own.

National 5 physics and equations

National 5 physics course

An experimental and investigative approach will be used as you undertake your course. The visual content of this revision guide reflects this and will help develop the knowledge and understanding of the concepts you require for your examination.

Studying this course helps us advance our understanding of the world around us and develop technological improvements which enhance all our lives.

The scientific inquiry skills you develop will help you discuss and review the many science based claims you meet in general literature.

This course lays the foundations for further study or for those entering many areas of work.

The National 5 physics course has six main areas. The topics covered in each of these areas of the course are shown below:

Dynamics

Scalars and vectors; velocity-time graphs; acceleration; Newton's laws; energy and projectile motion.

Space

Space exploration and cosmology.

Electricity

Electrical charge carriers; potential difference (voltage); Ohm's law; electrical and electronic circuits; electrical power.

Properties of matter

Specific heat capacity; specific latent heat; gas laws and the kinetic model.

Waves

Wave parameters and behaviours; electromagnetic spectrum and refraction of light.

Radiation

Nuclear radiation.

During your National 5 course you will also learn to use units, prefixes, significant figures and scientific notation correctly as seen on page 130–131.

Essential relationships

$d = vt$

$d = \bar{v}t$

$s = vt$

$s = \bar{v}t$

$a = \frac{v - u}{t}$

$F = ma$

$W = mg$

$E_w = Fd$

$E_p = mgh$

$E_k = \frac{1}{2}mv^2$

$Q = It$

$V = IR$

$V_2 = \left(\frac{R_2}{R_1 + R_2}\right)V_s$

$\frac{V_1}{V_2} = \frac{R_1}{R_2}$

$R_T = R_1 + R_2 + \ldots$

$\frac{1}{R_T} = \frac{1}{R_1} + \frac{1}{R_2} + \ldots$

$P = \frac{E}{t}$

$P = IV$

$P = I^2R$

$P = \frac{V^2}{R}$

$E_h = cm\Delta T$

$E_h = ml$

$P = \frac{F}{A}$

$p_1V_1 = p_2V_2$

$\frac{p_1}{T_1} = \frac{p_2}{T_2}$

$\frac{V_1}{T_1} = \frac{V_2}{T_2}$

$\frac{pV}{T} = \text{constant}$

$f = \frac{N}{t}$

$v = f\lambda$

$T = \frac{1}{f}$

$A = \frac{N}{t}$

$D = \frac{E}{m}$

$H = Dw_r$

$\dot{H} = \frac{H}{t}$

Additional relationships worth learning:

s = *area under a v-t graph*

a = *gradient of a v-t graph*

$v = u + at$

$$\frac{p_1 V_1}{T_1} = \frac{p_2 V_2}{T_2}$$

Series circuits:

$I_s = I_1 = I_2 = \ldots$

$V_s = V_1 + V_2 + \ldots$

Parallel circuits:

$I_p = I_1 + I_2 + \ldots$

$V_p = V_1 = V_2 = \ldots$

TOP TIP

Learn equations: it will save you time in the exams.

National 5 assessment

The assessment for the National 5 course has two parts: the exam question paper and the assignment.

You do not need to be concerned with this but the exam board will scale these parts into the ratio 80:20.

The course assessment is graded A - D based on the total of both parts.

Exam assessment

The duration of the exam question paper is 2 hours and 30 minutes.

This paper has a total marks of 135 (which will be scaled down to 100 marks).

You should only use blue or black ink, not gel pens.

You will gain most of the marks for demonstrating and applying knowledge and understanding. You will gain further marks for applying scientific inquiry and analytical thinking skills.

There are two sections in the question paper:

1. The physics objective test has 25 questions at 1 mark each. For each question you are provided with 5 answers only 1 of which is correct. You should indicate your choice on the answer grid which is in the paper.
2. Restricted (short) and extended response questions are in section 2. (These marks total 110 but these will be scaled to 75 so that the total for both parts of the question paper is 100). You are required to write your answers in the spaces provided in the question paper.

For a question that requires a calculation, it is good practice to write down the relationship you have chosen, the figures of the calculation and the final answer. If you have been asked to "show" or "prove" something then it is essential that all steps must be shown or you will get no marks.

Remember that a final answer to a calculation must include the relevant unit to gain the mark for the answer.

Example

Here is a guide to how to calculate resistance from a question.

1. Select the relationship: $V = IR$
2. Substitute the values: $7{\cdot}5 = 1{\cdot}5 \times R$
3. Answer with unit: $R = 5{\cdot}0\Omega$

TOP TIP

This question has 3 marks.

Remember that if you are asked to **explain**, **justify** or **suggest**, you are expected to give more than just a simple statement. You may want to consider the relationship between variables and how they interact.

TOP TIP

Remember to use the data and relationship sheets provided in the question paper.

Assignment

It is recommended that the duration of the assignment is 8 hours in total.

The assignment has a total of 20 marks (which will be scaled up to 25 marks).

Practical/experimental work is a mandatory feature of the assignment.

The assignment has two stages:

1. A **research** stage
2. A **report** stage

Research

Most of your time will be spent in this preparation stage.

- The experiment you choose must allow measurements to be made. For example you may want to investigate the relationship between two variables.
- The topic chosen and the experimental work should be at a level suitable for National 5 and agreed with your teacher or lecturer.
- You must also gather data from the internet, books and/or journals to compare against your own experimental results.

TOP TIP

At this stage check that your aim is suitable.

Report

You have 1 hour and 30 minutes to spend on this stage. This will be done under supervision. You should ensure you have:

- A copy of "Instructions for candidates"
- Your raw experimental data

- Internet or literature data (including a note of the sources used)
- Information you gathered on the underlying physics
- The experimental method, if appropriate

Your report must follow a clear structure:

- Title
- Aim
- Underlying physics
- Brief experimental method
- Experimental data (include all measurements, headings and calculate averages here)
- Graphical presentation (fully labelled)
- Research data (from internet/literature source, with reference)
- Analysis
- Conclusion
- Evaluation

There is no word count.

The report will be marked by the exam body.

TOP TIP

The conclusion should relate to your aim.

Specified data

Physics to learn	Identify data.
Revision Guide	You must use these values as provided on the exam paper Data Sheet.

Essential data

Speed of light in materials

Material	Speed (ms^{-1})
Air	$3{\cdot}0 \times 10^{8}$
Carbon dioxide	$3{\cdot}0 \times 10^{8}$
Diamond	$1{\cdot}2 \times 10^{8}$
Glass	$2{\cdot}0 \times 10^{8}$
Glycerol	$2{\cdot}1 \times 10^{8}$
Water	$2{\cdot}3 \times 10^{8}$

Gravitational field strengths

Location	Gravitational field strength on the surface (kg^{-1})
Earth	9·8
Jupiter	23
Mars	3·7
Mercury	3·7
Moon	1·6
Neptune	11
Saturn	9·0
Sun	270
Uranus	8·7
Venus	8·9

Specific latent heat of fusion of materials

Material	Specific latent heat of fusion (J kg^{-1})
Alcohol	$0{\cdot}99 \times 10^{5}$
Aluminium	$3{\cdot}95 \times 10^{5}$
Carbon dioxide	$1{\cdot}80 \times 10^{5}$
Copper	$2{\cdot}05 \times 10^{5}$
Iron	$2{\cdot}67 \times 10^{5}$
Lead	$0{\cdot}25 \times 10^{5}$
Water	$3{\cdot}34 \times 10^{5}$

Specific latent heat of vaporisation of materials

Material	Specific latent heat of vaporisation (J kg^{-1})
Alcohol	$11{\cdot}2 \times 10^{5}$
Carbon dioxide	$3{\cdot}77 \times 10^{5}$
Glycerol	$8{\cdot}30 \times 10^{5}$
Turpentine	$2{\cdot}90 \times 10^{5}$
Water	$22{\cdot}6 \times 10^{5}$

Specified data

Speed of sound in materials

Material	Speed (ms^{-1})
Aluminium	5200
Air	340
Bone	4100
Carbon dioxide	270
Glycerol	1900
Muscle	1600
Steel	5200
Tissue	1500
Water	1500

Specific heat capacity of materials

Material	Specific heat capacity ($J\,kg^{-1}\,°C^{-1}$)
Alcohol	2350
Aluminium	902
Copper	386
Glass	500
Ice	2100
Iron	480
Lead	128
Oil	2130
Water	4180

Melting and boiling points of materials

Material	Melting point (°C)	Boiling point (°C)
Alcohol	−98	65
Aluminium	660	2470
Copper	1077	2567
Glycerol	18	290
Lead	328	1737
Iron	1537	2737

Radiation weighting factors

Type of radiation	Radiation weighting factor
alpha	20
beta	1
fast neutrons	10
gamma	1
slow neutrons	3

National 5 PHYSICS

For SQA 2019 and beyond

Revision Guide

John Taylor

Scalars and vectors

Physics to learn	Definition of scalar and vector quantities.
Revision guide	Identification of terms, simple addition.

Scalars and vectors

We classify physical quantities as **scalar** or **vector**.

Scalars

A scalar quantity is defined by its magnitude alone. This means that it only has size. Scalar quantities are added using basic arithmetic.

Example

A lorry has a mass of 1500 kg and picks up a load of 500 kg. What is its new combined mass?

Just add them up!

1500 + 500 = 2000 kg

Vectors

A vector quantity is defined by both its magnitude and its direction.

Example

A basket and its load for a balloon flight have a weight of 5000 N. During the flight the balloon exerts a force up of 5000 N. What is the combined force?

If we just add the sizes we would obtain a value of 10 000 N. This is the wrong method! As one force is up and the other is down the forces should cancel each other out. The value will be 0 N so that the balloon neither rises nor descends.

TOP TIP

Magnitude is another name for size.
Below are some examples of scalar and vector quantities.

Scalar	Vector
distance	displacement
speed	velocity
mass	weight
time	force
energy	acceleration
power	momentum

Distance and displacement

Distance (d) is a scalar quantity.

It is defined by a number and its unit, the metre (m).

Displacement (s) is a vector quantity.

For example, a hiker walks 12 km due east from the village car park. Displacement is the distance travelled from a point in a certain direction. The direction may need some reference point, e.g. 30° east of north.

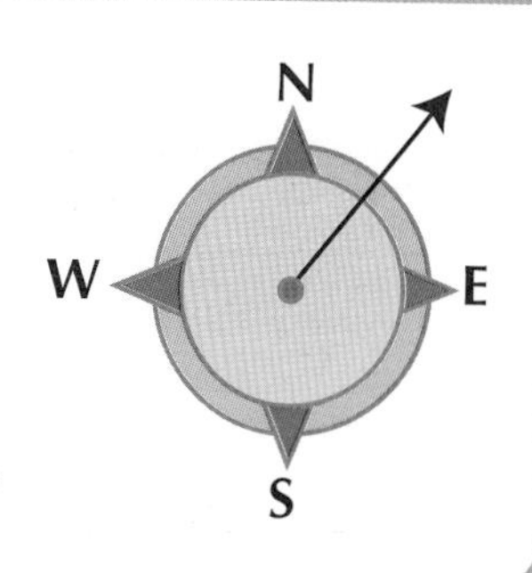

Speed and velocity

Speed is a scalar quantity.

The speed of an object is defined as the distance travelled per unit time (1 s).

Speed only has magnitude.

A speed limit of $20\,ms^{-1}$ does not need a direction.

Velocity is a vector quantity.

The velocity of an object is the speed in a given direction along a straight line. It is defined as the displacement per unit time (1 s).

Speed and velocity are measured in ms^{-1}.

A man walks along a conveyor belt in the opposite direction to the belt's movement. He is walking at a speed of $2\,ms^{-1}$. The conveyor belt is moving at a speed of $1{\cdot}5\,ms^{-1}$. The velocity of the man over the ground is only $0{\cdot}5\,ms^{-1}$ in the original direction.

Velocity is equivalent to speed with a given direction.

TOP TIP

We need to remember to state a direction when asked for displacement or velocity.

Quick Test 1

1. State what should be given to fully describe a vector quantity.
2. Give **two** examples of scalar quantities.
3. Give **two** examples of vector quantities.
4. Distinguish between distance and displacement.
5. Distinguish between speed and velocity.

Measuring displacement and velocity

Physics to learn	Displacement, average and instantaneous velocities.
Revision guide	You can describe experiments to measure these quantities.

Measuring distance and displacement

Distance and displacement can both be measured with a metre stick or measuring tape. With displacement we must also give a direction. Displacement is taken to be the shortest straight-line distance from start to finish in a certain direction.

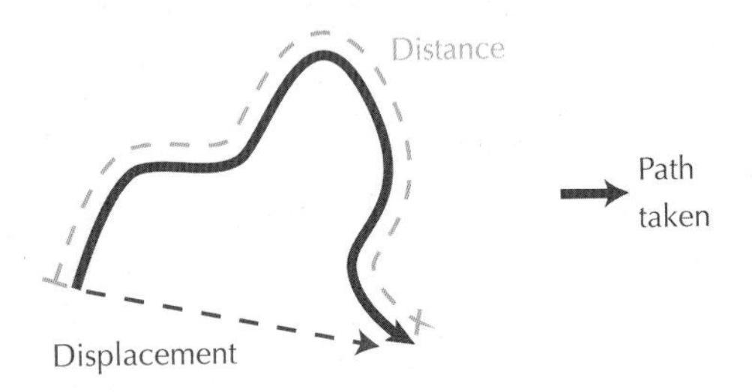

TOP TIP

The Latin word for distance is *spatium*, so we use s for displacement.

Measuring average velocity

Average speed is the total distance travelled over the total time taken.

$$\text{average speed} = \frac{\text{total distance}}{\text{total time}}$$

To measure the average speed of a cyclist on a road we could use a measuring tape and a stopwatch.

- Use the tape to measure a marked distance (e.g. 100 m).
- Use the stopwatch to measure the time taken (e.g. 6·50 s).
- Use the formula $v = \frac{d}{t}$ to calculate the average speed as $\frac{100}{6 \cdot 50} = 15 \cdot 4\ \text{ms}^{-1}$.
- Finally, to state the average velocity we need to note which direction the cyclist is travelling and add that to our description.

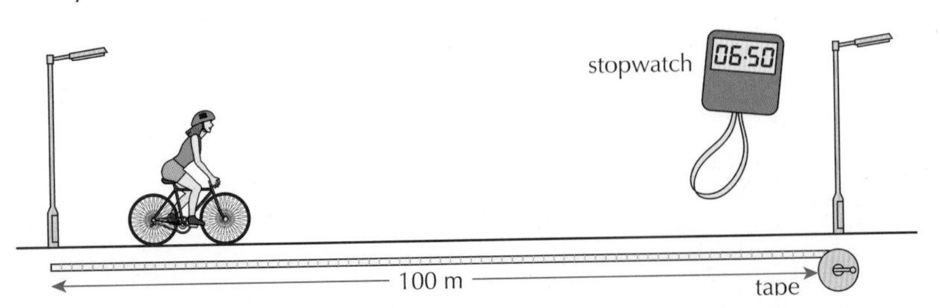

TOP TIP

to calculate distance $d = \bar{v}t$

Measuring instantaneous velocity

Instantaneous speed is the speed at a certain time. A good estimate of instantaneous speed is obtained by using a very small time interval.

To measure the speed of a toy car a light-gate is attached to an electronic timer or computer timer.

A short length of card is attached to the toy car. The length of card passes through the light beam as the car moves down the slope.

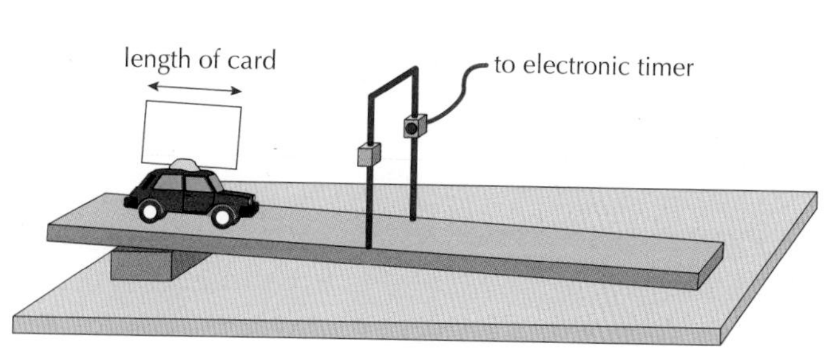

- Use a ruler to measure the length of card.
- Use the light gate and electronic timer to measure the short time taken by the card through the beam.
- Use the formula $v = \frac{d}{t}$ to calculate the instantaneous speed at the light gate.
- To state the instantaneous velocity we could give the speed and state the direction as 'down the slope' or a similar statement.

Distinguishing average and instantaneous velocities

TOP TIP

List situations where instantaneous and average speed are different.

During a car journey, at different times the car increases speed, changes direction, decreases speed and occasionally stops.

The instantaneous velocity has to be calculated repeatedly.

The average velocity only needs to be calculated once from the total displacement and the total time:

$$\text{average velocity} = \frac{\text{total displacement}}{\text{total time}} \qquad \bar{v} = \frac{d}{t}$$

The displacement is the journey in a straight line from start to finish.

Quick Test 2

1. Describe how to measure displacement.
2. Describe how to measure average velocity.
3. Describe how to measure instantaneous velocity.
4. Explain why instantaneous speed changes.
5. How could you mark the difference between forward and backward?

Resultant vectors

Physics to learn	Calculation of resultant vector.
Revision guide	Can use scale drawing and mathematical methods with vector addition.

A runner goes round a race track three times. If the track has a length of 400 m, what is the runner's distance travelled and displacement?

Distance travelled: d = 3 × 400 = 1200 m.

Displacement is how far he is from the starting position. As he has returned to the start, displacement: s = 0 m.

Scale drawings

A vector can be represented as a line drawn to scale with an arrow to show the direction. To combine vectors:

1. Choose and write down a scale.
2. Add vectors drawn 'head-to-tail'.
3. Draw the resultant from 'start to finish'.
4. Measure both the magnitude and the direction.

Example 1

A walker walks 1000 m east before returning to a cafe 700 m back west.

Following the rules above, we can show the:

- scale: 1 cm = 100 m
- distance travelled: d = 1000 + 700 = 1700 m
- and the displacement: 1000 + (–700) = 300 m east.

Example 2

A sailor sets his boat on a heading of north at 5 ms^{-1} through the sea.

The tide is moving at 2 ms^{-1} in an easterly direction.

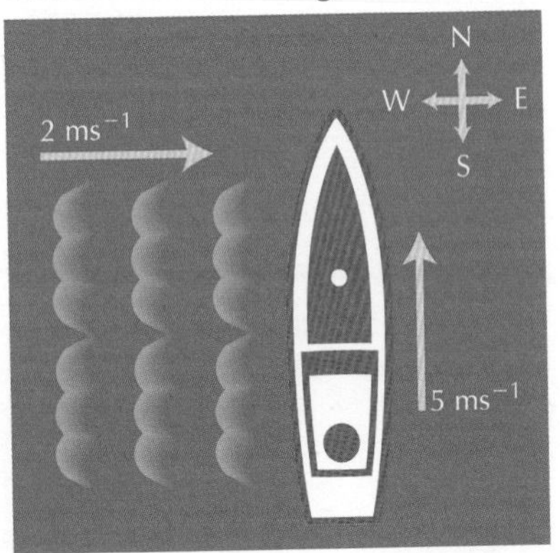

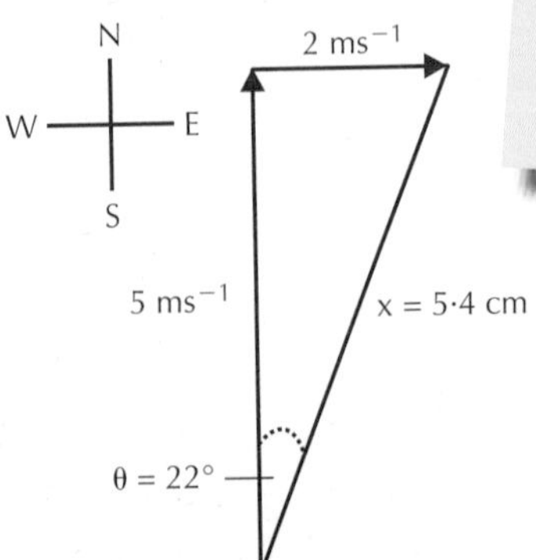

TOP TIP

If a bearing has been used for the direction, 000° is north.

TOP TIP

Do not forget the direction – it's part of the vector, so you could lose marks!

Find the boat's resultant velocity. As the vectors will be at right angles, you will have the option of using either a scale diagram or basic trigonometry.

1. Scale: 1 cm = 1 s^{-1}.
2. Boat and tide velocities added on head-to-tail diagram.
3. Resultant drawn start to finish.
4. a) Measured resultant, x = 5·4 cm.
   b) Measured resultant angle, θ = 22°.

Resultant velocity, v = 5·4 ms^{-1} at 022°.

Trigonometry

The mathematical rules

For size, use the Pythagorean theorem: $c^2 = a^2 + b^2$.

For direction, use trigonometric methods:

- cos θ = adjacent/hypotenuse
- sin θ = opposite/hypotenuse
- tan θ = opposite/adjacent.

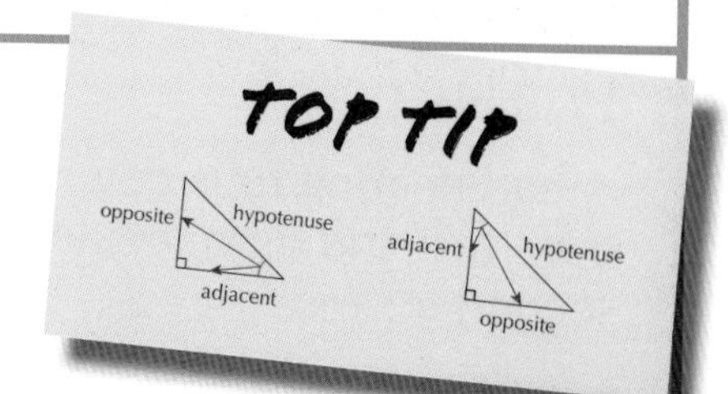

Example

A car drives 4 km north then drives 3 km west. What is the resultant displacement?

Using Pythagoras and trig:

$$c^2 = a^2 + b^2$$
$$x^2 = 4^2 + 3^2 = 16 + 9$$
$$x^2 = 25$$
$$x = 5\,\text{km}$$
$$\tan\theta = \frac{3}{4}$$
$$\theta = \tan^{-1}\frac{3}{4}$$
$$\theta = 37^\circ$$

displacement (s) = 5 km at 37° west of north. You can use a ruler and protractor to check this by scale measurement.

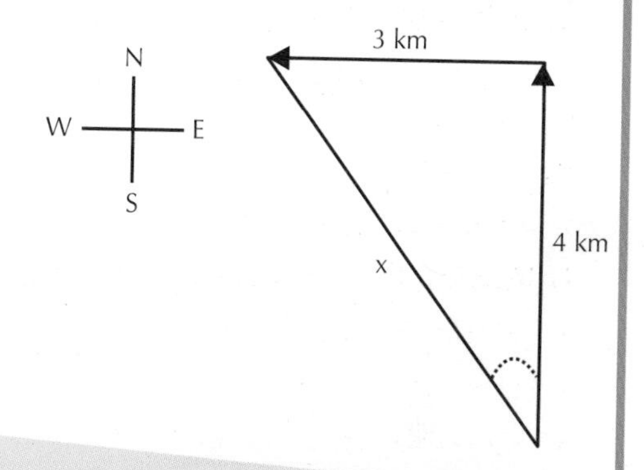

TOP TIP

tan θ = $\frac{3}{4}$ means you need to use the inverse function to find the angle. On your calculator find $\tan^{-1}$ or Inv tan. If you're getting, for instance, 0·01309 or 76·390 as your answer, you're not using your calculator properly!

Quick Test 3

1. State what should be given to fully describe a vector quantity.
2. A force of 90 N is exerted on a weight along a bench and a force of 30 N is exerted on the weight across the bench. Calculate the resultant force being exerted.
3. Two tugs pull on a boat with a force of 2 kN each and an angle between them of 90°. Calculate the size of the resultant force being exerted.

Velocity-time graphs

Physics to learn	From data, draw velocity-time graphs.
Revision guide	Can plot and interpret velocity-time graphs.

Plotting a velocity-time graph

TOP TIP

When recording data in a table, use headings that include the units (e.g. s, ms^{-1}) so that the data in the table are just the numbers.

Record the velocity of a moving object at regular time intervals.

Example 1

A ball bearing rolls down a grooved track from rest and its velocity is recorded every second for 6 s.

Time (s)	0	1	2	3	4	5	6
Velocity (ms^{-1})	0	0·1	0·2	0·3	0·4	0·5	0·6

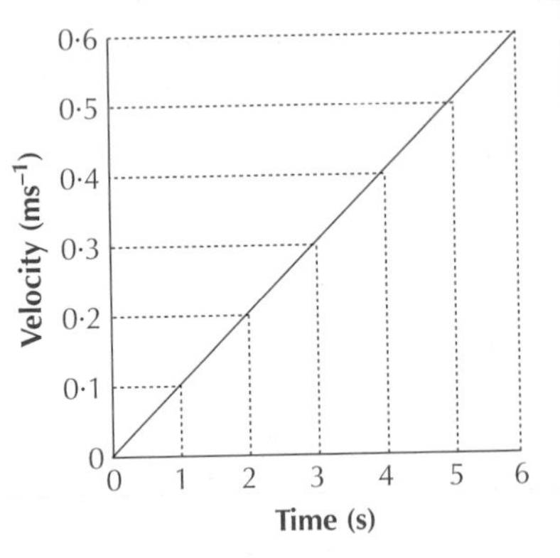

Example 2

A football manager paces along the touchline for 6 s at 2 ms^{-1} before taking 1 s to stop and turn around then returning at 1 ms^{-1} for 8 s.

Time(s)	0	1	2	3	4	5	6	7	8	9	10	11	12	13	14	15
Velocity(ms^{-1})	2	2	2	2	2	2	2	−1	−1	−1	−1	−1	−1	−1	−1	−1

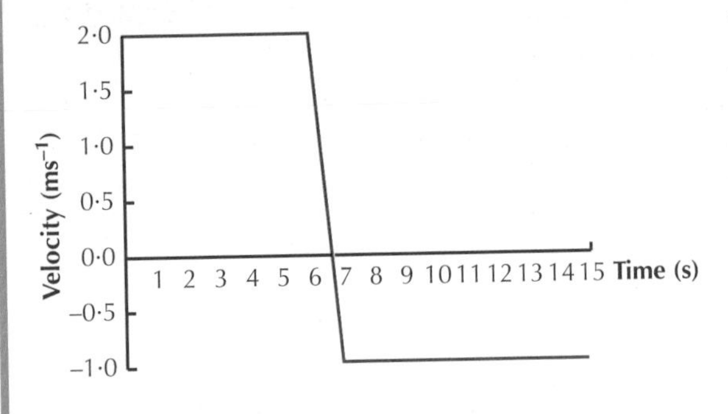

Why do the values become negative for the manager's walk?

As there is a change of direction we take forward to be positive and backward to be negative.

TOP TIP

In these graphs, time should always be plotted on the x (horizontal) axis. The velocity is the variable factor and should go on the y (vertical) axis.

Interpreting graphs

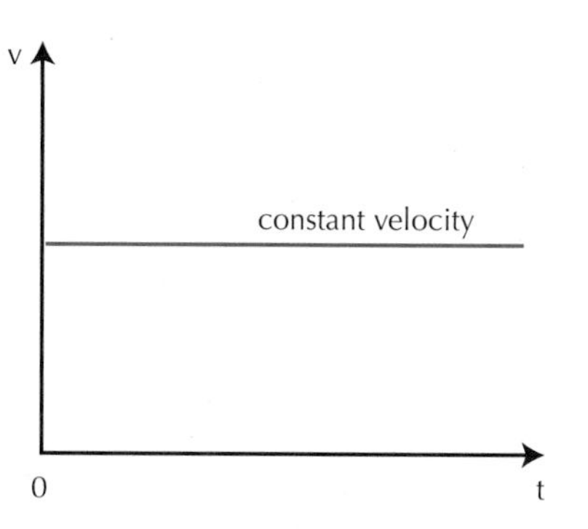

Horizontal line: object moving at constant velocity.

TOP TIP

When drawing a graph remember to:

- use a sharp pencil
- use a ruler for straight lines and axes
- label the axes and include units.

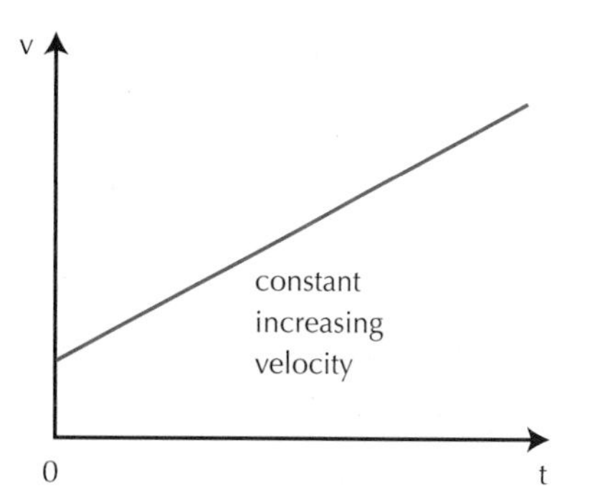

Straight line sloping upwards: object increasing speed: constant acceleration (positive).

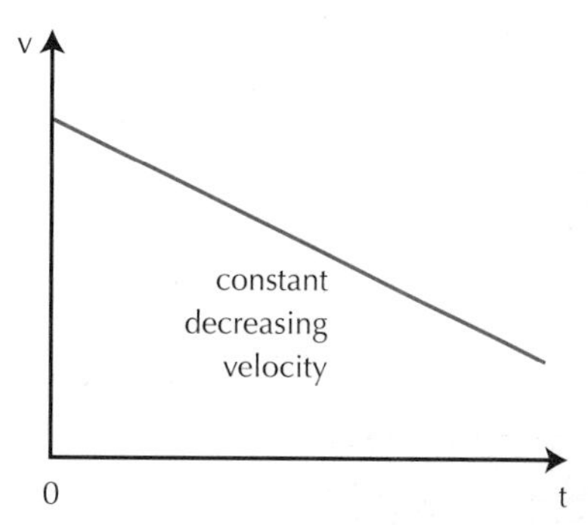

Straight line sloping downwards: object decreasing velocity: constant acceleration (negative).

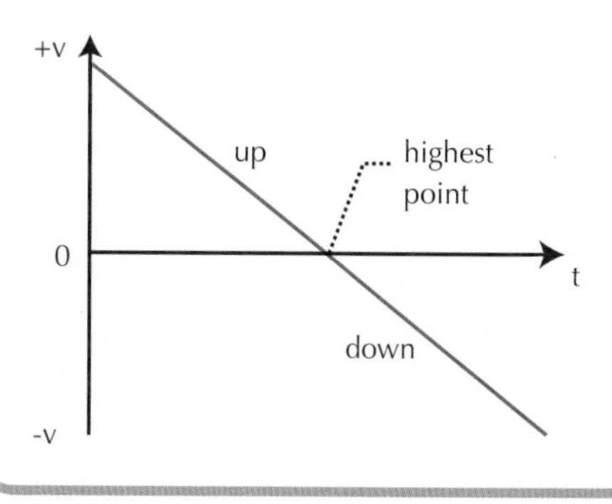

Straight line crossing axis, e.g. a ball thrown up in the air and returns: object has changed direction. The speed decreases then increases in the opposite direction. The acceleration is constant and negative.

Quick Test 4

1. What is the independent variable and where does it go?
2. What is the dependent variable and where does it go?
3. State what a change of axis signifies.
4. State what sensor can be used for time and velocity.

Velocity and displacement

Physics to learn	Displacement from a velocity-time graph.
Revision guide	Can use a velocity-time graph to find an object's displacement.

Calculating displacement

Displacement is the distance travelled in a straight line from start to finish. Displacement is related to the average velocity by:

$$s = \bar{v}t$$

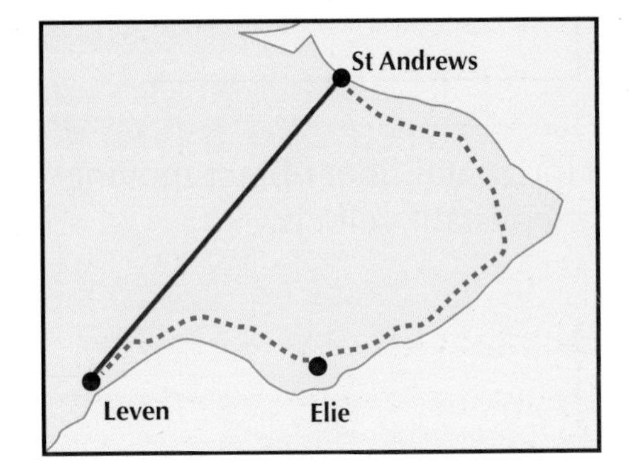

Travelling at constant velocity

A plane is flying at a constant velocity of $200\,ms^{-1}$ at 110° for 1 h 40 min.

The displacement can be calculated by formula:

$$s = vt = 200 \times (100 \times 60) = 1\,200\,000\,m$$

displacement = 1 200 km at 110°

The journey can be plotted on a v-t graph. The area under the v-t graph is the product of velocity on the *y* axis and time on the *x* axis.

Area = 200 × 6000 = 1 200 000 m

The size of the displacement at 1 200 km is the same.

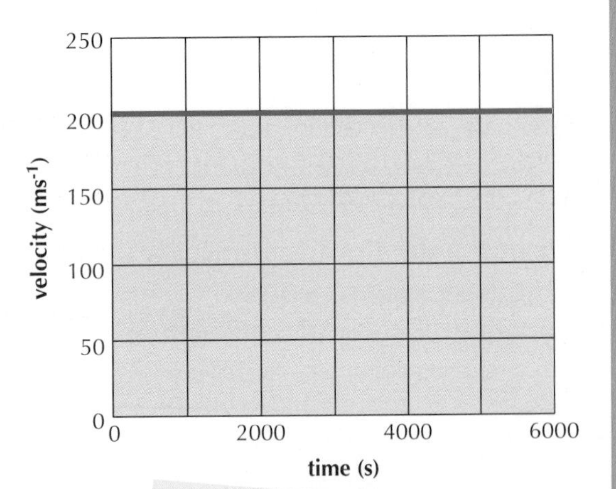

TOP TIP

The rectangular area is equivalent to l × b.

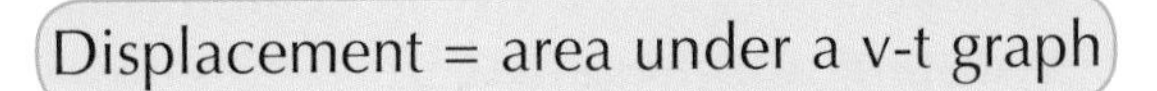

We now have several alternative equations to use for displacement:

- $s = vt$
- $s = \bar{v}t$
- s = area under a v-t graph

We choose the equation according to the relevant circumstance.

Displacement when accelerating

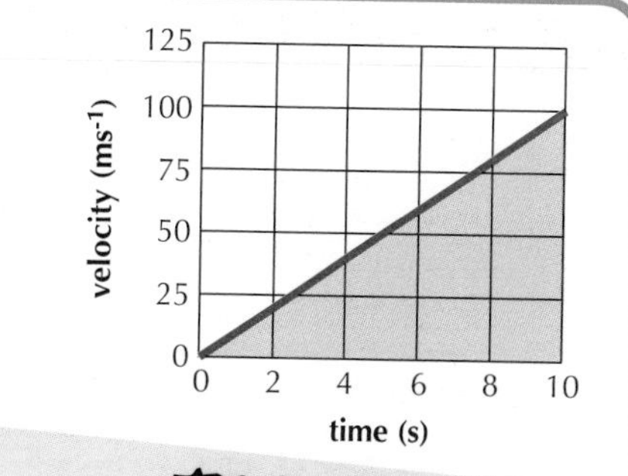

A racing car accelerates from rest to 100 ms^{-1} in 10 s. Assuming it is constantly accelerating, draw a velocity-time graph and calculate the displacement required for this motion.

The equation:

displacement = area under a v-t graph,

can be applied when the motion is not constant. This is a useful alternative to the simple equation for displacement.

Displacement = area under a v-t graph

$$= (\frac{1}{2}(100 - 0) \times 10)$$

$$= 500\,\text{m}$$

TOP TIP

The triangular area is equivalent to $\frac{1}{2} \times l \times b$.

Displacement on a journey

The type of motion on a journey varies. Divide the velocity-time graph into different areas. The total displacement is the sum of each area.

Example

A motorcyclist starting from rest accelerates to a speed of 40 ms^{-1} in 4 s. He travels at this speed for 10 s before decelerating to a halt in 8 s.

Displacement = area under a v-t graph

$$= (\frac{1}{2} \times 40 \times 4) + (40 \times 10) + (\frac{1}{2} \times 40 \times 8)$$

$$= 640\,\text{m}$$

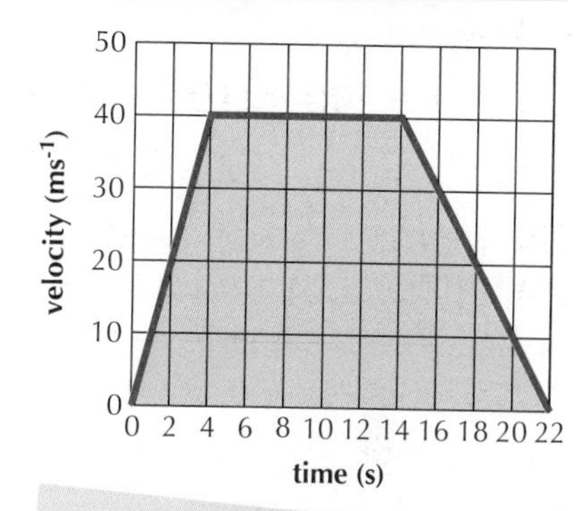

TOP TIP

If the velocity-time graph goes negative, the displacement is returning towards the start.

Quick Test 5

1. A car has to do an emergency stop. The car keeps constant velocity during reaction time. The brakes then operate to decelerate the car to rest.

   The car was going at 20 ms^{-1}, reaction time took 0·7 s and the car took a further 2 s to come to rest.

   Calculate the distance the car took to stop.
2. A cyclist cycles steadily at 5 ms^{-1} for 10 minutes from Glasgow towards Edinburgh. Calculate her displacement.

Acceleration

Physics to learn	Acceleration.
Revision guide	Define, calculate and measure acceleration.

Describing acceleration

Acceleration (a) is the change of velocity in unit time (1 s). Acceleration is also defined as the rate of change of velocity.

$$\text{acceleration} = \frac{\text{final velocity} - \text{initial velocity}}{\text{time}}$$

$$a = \frac{v - u}{t} \qquad v = u + at$$

What is deceleration?

Acceleration is a vector. When the final velocity is less than the initial velocity, the acceleration will be negative.

A negative acceleration is a deceleration.

TOP TIP

For deceleration, use the acceleration equation. The answer is negative. An acceleration of $-5\,ms^{-2}$ is a deceleration of $5\,ms^{-2}$.

Example

A dog slows from $12\,ms^{-1}$ to rest in 4 s. What is its deceleration?

$$a = \frac{v - u}{t}$$

$$a = \frac{0 - 12}{4} = -3\,ms^{-2}$$

The deceleration is $3\,ms^{-2}$.

Measuring acceleration

Acceleration on a slope

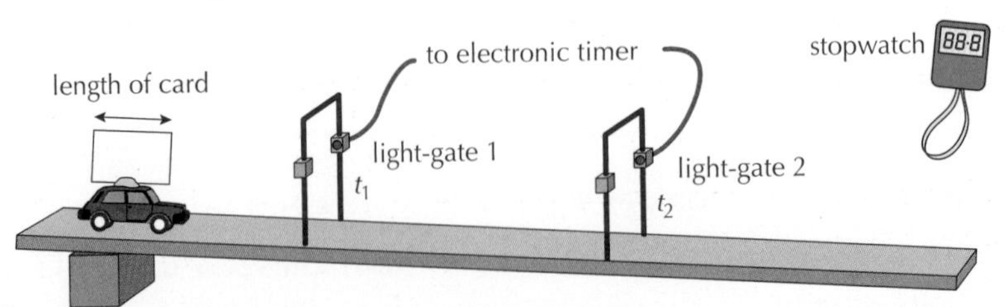

We can measure acceleration using two light gates and a stopwatch. We must obtain the velocity at two points on the slope and the time interval between them. A short measured length of card is attached to the vehicle to cut the light beam.

- At light gate 1 we use the length of card through light gate 1 and time t_1 to obtain an initial velocity, u. $u = \frac{d}{t_1}$
- At light gate 2 we use the length of card through light gate 2 and time t_2 to obtain a final velocity, v. $v = \frac{d}{t_2}$
- The stopwatch is used to record the time, t, between these two velocities.
- Then we can use the equation $a = \frac{v - u}{t}$ to calculate the average acceleration.

Acceleration at a point

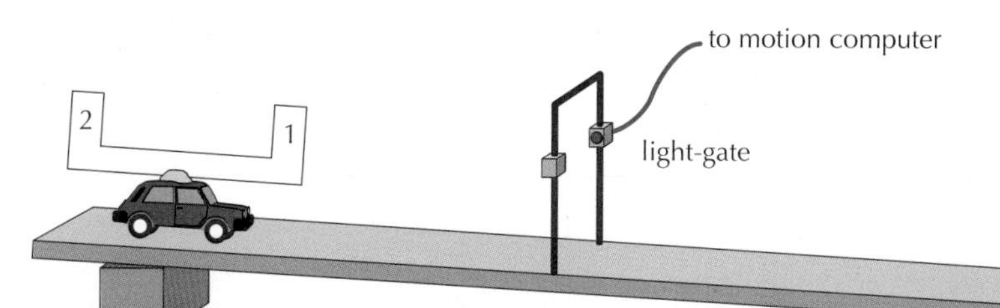

TOP TIP

The computer has been programmed with the velocity and acceleration equations.

We can also measure acceleration using two cards and one light gate attached to a motion computer. This will find the acceleration at one point on the track. The second card cuts the light beam quicker than the first. The length of card must be measured and entered into the motion computer.

- Card 1 is timed and the computer calculates velocity, u, using the length of the card and time t_1.
- Card 2 is timed and the computer calculates velocity, v, using the length of the card and time t_2.
- The motion computer also records the time between the two cards cutting the light beam and then calculates the acceleration, a.

Quick Test 6

1. State what is meant by the term acceleration.
2. What is a negative acceleration?
3. State what quantities need to be measured to calculate acceleration.
4. Calculate the acceleration of a car that increases its speed by 60 ms^{-1} in 20 s.
5. A VW is travelling at 10 ms^{-1} when it accelerates at 5 ms^{-2} for 3 s. What is its new speed?
6. A bus is travelling at 20 ms^{-1} when it decelerates at 2 ms^{-2}. How long does it take to stop?
7. A train slows from 35 ms^{-1} to 20 ms^{-1} in 5 s. Calculate its acceleration and deceleration.
8. Explain which method above would give the least accurate measurement of acceleration.

Acceleration graphs

Physics to learn	Calculating acceleration and gradients.
Revision guide	Determination of acceleration from a velocity-time graph.

Calculating acceleration

Calculating acceleration from a velocity-time graph

We can find the acceleration of an object by marking two points on a velocity-time graph and extracting data for the acceleration equation.

- Measure the change in velocity, $v - u$, from the y axis.
- Measure the change in time, t, from the x axis.

$v - u = 30 - 0$

$t = 5 - 0$

$$a = \frac{v-u}{t} = \frac{30-0}{5-0} = 6 \text{ ms}^{-2}$$

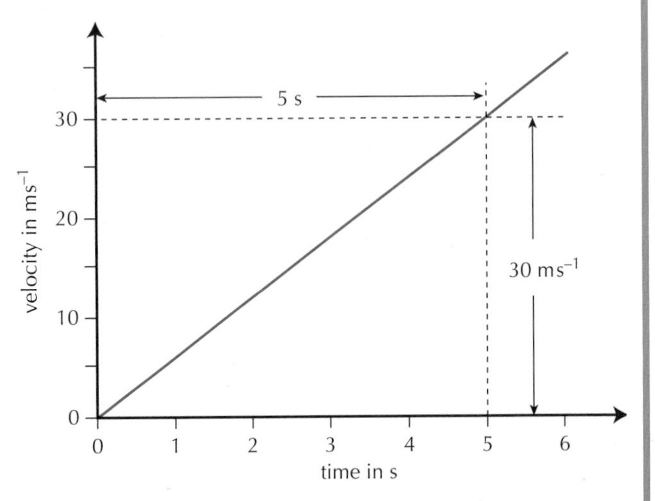

TOP TIP

a = gradient of the line on a v-t graph

Gradients

Calculating acceleration from the gradient of a v-t graph

We can find the acceleration of an object by measuring the gradient of its velocity-time (or speed-time) graph.

$y_2 - y_1 = v - u$

$x_2 - x_1 = t$

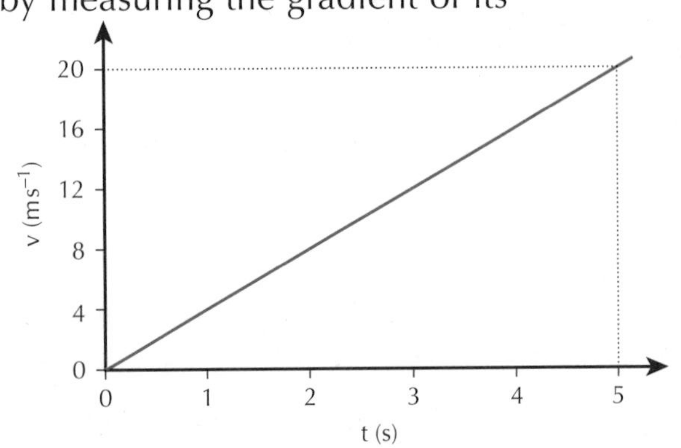

Example 1

The acceleration of the object in the graph is:

$$a = \frac{v-u}{t} = \frac{20-0}{5-0} = 4 \text{ms}^{-2}$$

Example 2

A motorcyclist starting from rest accelerates to a speed of $40\,ms^{-1}$ in 4 s. He travels at this speed for 10 s before decelerating to a halt in 8 s.

Check:

0 – 4 s: $a_1 = 10\,ms^{-2}$

4 – 14 s: $a_2 = 0\,ms^{-2}$

14 – 22 s: $a_3 = -5\,ms^{-2}$

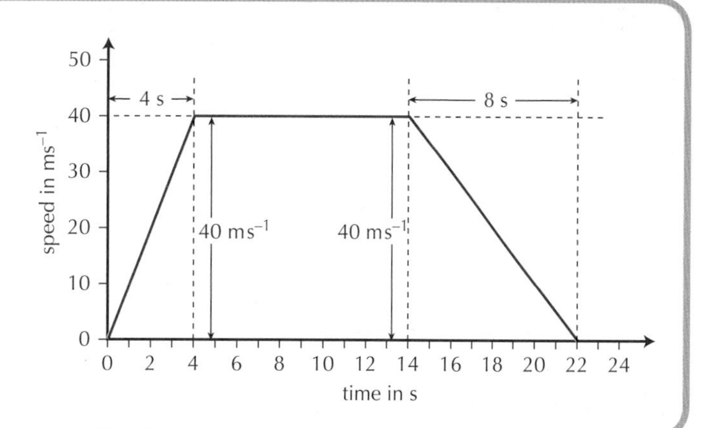

Identifying gradients

We can describe the acceleration from the gradient of a velocity-time graph.

On a velocity-time graph, if the line is horizontal and on the x axis then the velocity is zero and the object is stationary.

TOP TIP

If the gradient is negative, the acceleration is negative!

TOP TIP

The largest acceleration is calculated from the line with the steepest gradient.

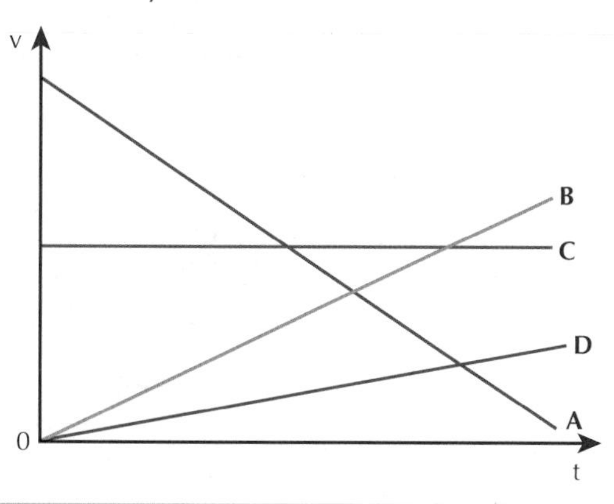

A = *constant deceleration*
B = *constant acceleration*
C = *zero acceleration*
C = *constant velocity*
D = *constant acceleration*
Note: the acceleration of B is greater than the acceleration of D.

Quick Test 7

1. State what the following show on a speed-time graph:
   (a) a horizontal line
   (b) a straight line sloping steeply upwards
   (c) a straight line sloping gently downwards.
2. Draw a speed-time graph to describe the following journey:
   A sprinter starting from rest accelerates to a speed of $10\,ms^{-1}$ in 2 s.
   He travels at this speed for the next 8 s then decelerates to $2\,ms^{-1}$ in 4 s.
   He continues to jog at this speed for the next 6 s.

Newton's 1st law of motion

Physics to learn	Newton's first law.
Revision guide	Application of the first law and balanced forces to explain constant velocity.

Isaac Newton

In 1687, Isaac Newton's book *Philosophiae Naturalis Principia Mathematica* or *Mathematical Principles of Natural Philosophy* made him one of the most important physicists of all time. As well as Newton providing important contributions to optics and calculus, his ideas on gravity and three laws of motion can still be used to explain the motion of most objects today.

Newton's first law

- An object at rest will remain stationary unless acted on by an unbalanced force.
- An object in motion will continue at the same speed in the same direction unless acted on by an unbalanced force.

Example

An aircraft is flying at a constant velocity.

- Constant speed: size of driving force = size of air resistance.
- Constant altitude: size of lift force = size of weight.

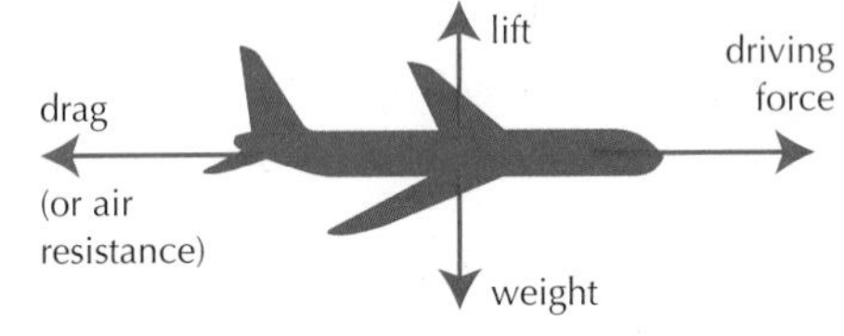

Newton's first law (N1): An object will remain at rest or will remain at constant velocity unless acted on by an unbalanced force.

Newton's first law applies when there are no applied forces or the forces are balanced.

The motion remains the same.

Question: What should you do to keep an object moving?

Answer: Do nothing! An object should just keep moving, see N1.

On Earth we expect objects to stop. Newton says something must be doing the stopping. This is the force of friction acting between the object and the ground. Without friction the object would just keep going.

In space flight we can see the benefit of Newton's first law. Spacecraft cannot carry fuel for more than a few minutes of burn. In space, without friction, we can say there are no forces applied and objects will keep going without energy being used or the rockets being on.

Stationary motion

The forces on the book are balanced
The table pushes upward on the book.
PHYSICS
Gravity pulls downward on the book.

The book on the table could be said to be 'at rest'. Newton said there must be no unbalanced forces acting. We can see that there are forces acting but they are balanced. This is the same as no forces acting. Newton's first law says the objects will remain at rest.

Advanced thinking

Newton also knew that objects move with the Earth's rotation. So are 'constant velocity' and 'at rest' the same thing? Newton's first law says that whether you think the object is 'at rest' or at 'constant velocity', the object will keep its motion.

Constant velocity

TOP TIP

- Don't forget that force is a vector.
- Balanced forces and no force give the same effect.
- The force of friction increases with velocity.

To keep a vehicle moving, we normally have to keep applying a force. Why is this?

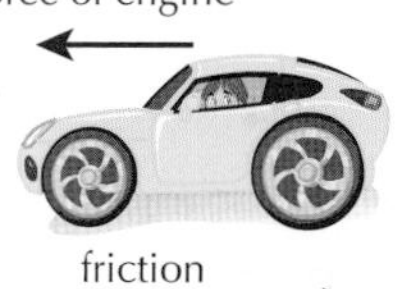

An object in motion experiences resistive forces that increase with velocity. These are known as the forces of friction.

Friction always acts against the direction of the motion.

Where the applied force and friction are balanced, Newton's first law tells us that the object will remain at constant velocity.

If the car is travelling at a higher speed then the engine force needs to be greater because the resistive friction forces have increased. The forces balance out again.

Frictionless motion

Friction is all around us and is often very useful. However, we may want to reduce friction to save energy loss. Here are some ideas to think about:

- The streamlined shape of a boat through the water reduces friction.
- A linear air track allows vehicles to float on a cushion of air in low friction experiments.
- The Maglev train might provide more transport for the future. Find out how the train uses repelling magnetic fields.

Quick Test 8

1. Describe the effect balanced forces have on the motion of an object.
2. Name a vehicle that keeps moving without requiring a force.
3. What outcome does Newton's first law have?
4. Describe what happens in a crash if you do not wear a seatbelt.
5. Why do we apply a force to keep an object moving?
6. What reduces friction for the Maglev train?

Newton's 2nd law of motion

Physics to learn	Newton's second law.
Revision guide	Application of the second law and unbalanced forces to explain or determine acceleration.

Acceleration and force

Unbalanced forces

The resultant of forces that do not cancel out is known as unbalanced force.

An unbalanced force causes acceleration. For instance, a racing car accelerates because its engine thrust is greater than the resistive forces. The difference in the opposing forces must be calculated.

Force, mass and acceleration

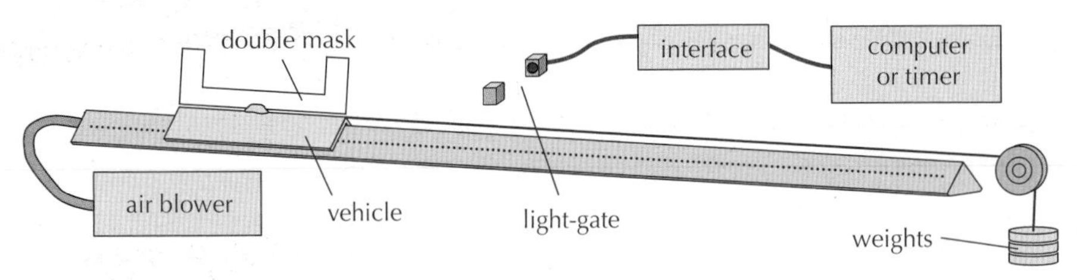

The air track reduces friction so it is not significant. The weights are applying a force to the moving system.

Experiment 1

The number of weights can be increased. The larger the force applied, the greater the acceleration. The acceleration varies with the unbalanced force.

This is called direct proportion: $a \propto F_{un}$

Experiment 2

The vehicle can have different masses. The larger the mass, the smaller the acceleration. The acceleration varies inversely with the unbalanced force.

This is called inverse proportion. $a \propto \frac{1}{m}$

Combining these conclusions: $a \propto \frac{F}{m}$

With mass in kg, acceleration in ms^{-2} and force in Newtons, this becomes:

$a = \frac{F}{m}$ or $F = ma$

TOP TIP

$F = ma$ also leads to $W = mg$ (see later)

Newton's second law

Newton's second law (N2): The acceleration of an object varies directly with the unbalanced force and inversely with its mass.

The newton is defined as the force that causes a mass of 1 kg to accelerate at 1 ms^{-2}. Newton's second law tells us that an unbalanced force causes acceleration.

TOP TIP

Unbalanced force can be written as F_{un}

Example

An 800 kg car is on a gentle slope. The slope is just enough to balance out the effect of friction. The driver starts the engine and it exerts a force of 2·4 kN. Calculate the car's acceleration.

$$a = \frac{F}{m} = \frac{2400}{800} = 3\,\mathrm{m\,s^{-2}}$$

TOP TIP

Balanced forces are equivalent to no force and cause no change in speed or direction. Unbalanced forces cause change.

Example

A 900 kg vehicle accelerates. The engine force is 2 kN, but friction exerts 200 N. Find the acceleration.

$$a = \frac{F}{m} = \frac{2000 - 200}{900} = 2\,\mathrm{m\,s^{-2}}$$

TOP TIP

Unbalanced force = applied force – force against friction.

Rocket launch

When a rocket launches:

- Liquid fuel engines (below the large central tank) are fired first.
- Thrust < weight. The rocket cannot accelerate or take off.
- Final checks are made and if there is a problem these engines can be shut down.
- If everything is OK, the solid fuel engines (the tanks on either side) are fired.
- Once ignited, these cannot be shut down and there is no going back!
- Thrust > weight. The rocket must take off, and it accelerates.

Quick Test 9

1. What quantity increases acceleration?
2. What quantity decreases acceleration?
3. How can we find the size of an unbalanced force?
4. A 3500 kg boat accelerates at 0·5 ms^{-2}. Calculate the unbalanced force.
5. A toy car of mass 500 g accelerates at 0·6 ms^{-2}. If the friction is 0·8 N, what is the size of the pulling force?
6. A weight has a mass of 0·5 kg. Calculate the force it exerts on earth.

Forces in action

Physics to learn	Balanced and unbalanced forces.
Revision guide	You can use free body diagrams and explain the effect of friction.

Balanced forces

balanced forces: no motion

Force is a vector quantity. Force has size and direction.

Forces opposing each other may cancel out.
Forces that cancel out are balanced forces and there is no change in motion.

Balanced forces are of equal size but opposite in direction.

Unbalanced forces

thrust (T)
weight (W)

Forces applied to an object that do not cancel out are unbalanced.

Unbalanced forces cause acceleration.

The rocket experiences thrust from the engines but has weight from the gravitational field. Only if the thrust is larger than the weight will the rocket take off and accelerate.

Free body diagrams

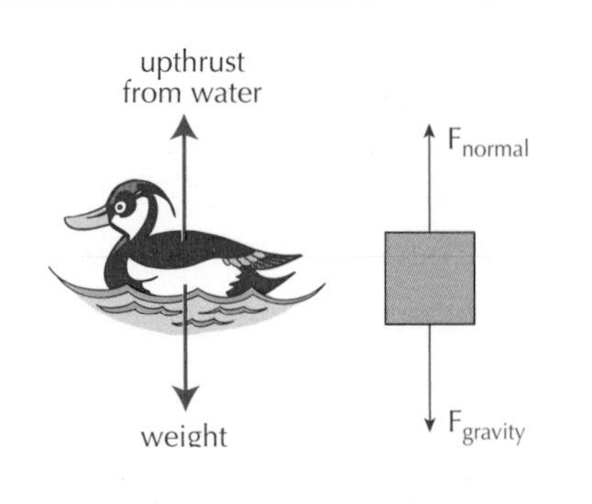

Free body diagrams can be used to show the forces acting on an object. A simple diagram is drawn and arrows are used to indicate the force and its direction. If one force is larger the arrow can be made longer to indicate this. The duck has weight, which acts downwards. The duck is not moving or sinking. The upthrust from the water must balance the weight.

TOP TIP

Weight down = the force of gravity down.

Friction

Friction is a resistance force that opposes motion. When two surfaces run together, small collisions occur, which creates a force of friction. Friction increases as the velocity increases and the surface area increases.

Investigating the force of friction

Pull a shoe along at a steady speed. The pull has the same strength as the friction. A rougher surface would increase frictional force. A heavier weight would increase the downward contact force and therefore the frictional force.

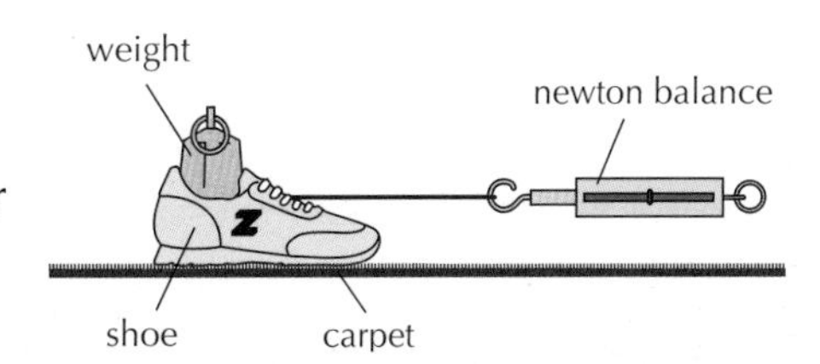

Increasing friction	Reducing friction
Increasing friction is useful to help an object speed up or slow down.  • Dry tyres for racing cars have no tread to increase the contact area with the road. • Aerofoils push cars down at speed for better grip in the corners. • Rough tarmac is laid before traffic lights. • A parachute has a large surface area for more air to hit.	Friction should be reduced where it opposes desired motion. • Streamlining of cars, cyclists, planes and bobsleighs reduces friction and saves fuel. • Wheels, rollers and ball bearings are inventions designed to reduce friction and allow movement – roll not slide! • Hovercrafts use a cushion of air to reduce contact with the ground. • Machinery is lubricated. This reduces the contact between the moving parts.

TOP TIP

Normal force is the force upwards from a surface.

Quick Test 10

1. Complete the following: Balanced forces are equal in ____________ but opposite in ____________.
2. Complete the following: Unbalanced force causes ____________.
3. A book is at rest on a table top. Draw a diagram of the forces acting on the book.
4. A hammer falls to the surface of the moon. Draw its free-body diagram.
5. Name **two** methods of reducing friction.

Newton's 3rd law of motion

Physics to learn	Newton's third law.
Revision guide	Identify Newton pairs and apply the third law.

Newton pairs

In Newton's second law we saw the effect of a force on an object. But Newton noticed that when there is a force in action there is also another force in the opposite direction. We call these forces 'Newton pairs.'

Forces occur in equal and opposite pairs.

You push on a wall and the wall pushes you away.

A Newton pair of forces has the following properties:

- These forces act on two different bodies at the same time.
- Both forces are of the same type (gravitational/electrostatic/magnetic).
- The forces are equal in magnitude.
- The forces are in opposite directions.
- The forces occur at exactly the same time, i.e. both objects will accelerate in opposite directions at the same time.

Identifying Newton pairs

Here are various situations where you should be able to identify Newton pairs. For each diagram, try to write down the names of the forces involved.

TOP TIP

Two force sensors measure the forces. Can they have different values? No, if one person increases their force, both will experience an increased force.

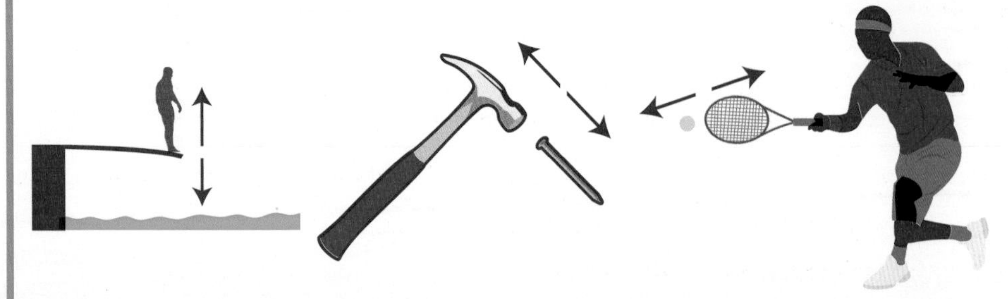

Complete this sentence for each of the Newton pair diagrams.

The _____ exerts a force on the _____; the _____ exerts a force on the _____

Newton's third law

Newton said the forces of two bodies on each other are always equal in size but are directed in opposite directions.

TOP TIP

N3 tells us that forces exist in pairs.

Newton's third law(N3): If A exerts a force on B, then B exerts an equal but opposite force on A.

TOP TIP

Newton called the pair of forces "action" and "reaction" but we should note that these occur only together and at the same time.

Newton's three laws and space travel

Newton's laws are used in space travel.

N3 tells us that forces exist in pairs: if the rocket exerts a force on the fuel, then the fuel exerts an equal but opposite force on the rocket.

Thus, the fuel pushing on the rocket causes acceleration. N2 tells us that a force causes acceleration.

N1 tells us we need no force for an object to keep going. An object will remain at rest, or stay at constant speed in a straight line, unless acted on by an unbalanced force. Interplanetary flight takes place at a constant speed using no fuel.

TOP TIP

Newton's laws describe virtually all motion.

Quick Test 11

1. Use N3 to describe the Newton pairs when a toy water rocket is blown up and let go.
2. Describe the initial motion of the water rocket and what law applies.
3. What motion does the rocket then develop and why?
4. A girl pushes herself out of a boat of mass 160 kg, with a force of 400 N. Her weight is 600 N. Calculate the initial acceleration of the boat.

Newton's laws in action

Physics to learn	Applications of Newton's laws.
Revision guide	You can apply the laws to car safety and terminal velocity.

Car safety

Seat belts and airbags

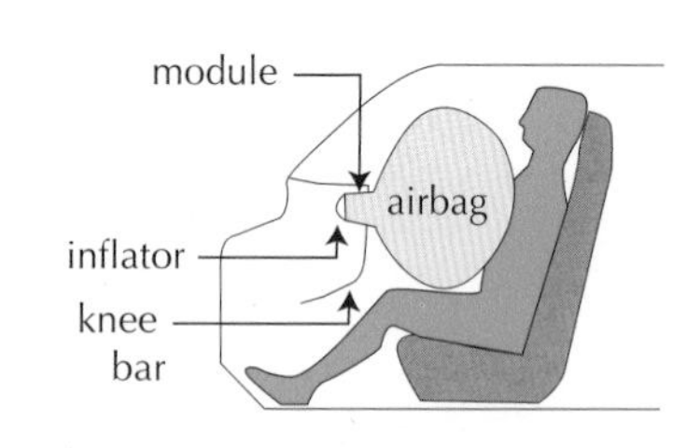

Consider objects in motion. N1 says that unless an unbalanced force acts, an object should keep its velocity. In a crash, if we are standing on a bus or do not have a seat belt on, Newton's first law says if no force is applied to us, we will just keep going forward at a constant velocity. We are not 'thrown forward'.

N2 shows that a seat belt applies a force in the opposite direction of motion, which decelerates the person with the vehicle. A seat belt will also have some 'give' so as not to cause injury.

A crash applies a large force in a short time. An airbag is designed to provide a longer time combined with a smaller force. The longer time is still only a fraction of a second but it can save your life.

TOP TIP

A small force for a long time has the same effect as a large force for a short time.

Stopping a car

To stop a car, applying the brakes makes the tyres exert a force on the road again, and the road exerts an opposite force on the tyres, stopping the car. Without Newton's third law the car would not stop safely!

Terminal velocity

1. A skydiver initially accelerates downwards at $9{\cdot}8\,ms^{-2}$ as the only downwards force is her weight.
2. As the skydiver's velocity increases, the air resistance increases and the acceleration is less. At a certain point in time she experiences air resistance of 600 N upwards. If her mass is 70 kg, determine her motion at that point.

   Resultant force, F_{un} = total downwards force – total upwards force
   = 686 – 600 = 86 N downwards

   $$\text{Acceleration, } a = \frac{F}{m} = \frac{686 - 600}{70} = 1{\cdot}2\,ms^{-2}$$

3. Air resistance increases until it balances weight and a final constant velocity is reached (at about 60 ms^{-1} without a parachute).

The final constant velocity of a moving object is known as **terminal velocity**.

When the skydiver opens the parachute, the greater surface area means a new lower terminal velocity will give the same air resistance to balance the weight, allowing the skydiver to descend safely.

TOP TIP

Air friction or resistance increases:

- as the velocity increases and
- as the frontal surface area increases.

Terminal velocity of a vehicle

TOP TIP

Better aerodynamics increases terminal velocity.

A. The driver presses the accelerator. The driving force makes the car accelerate.

B. The accelerator is kept down. Increasing air resistance => a smaller acceleration.

C. The accelerator is kept down. Air resistance now balances the driving force.

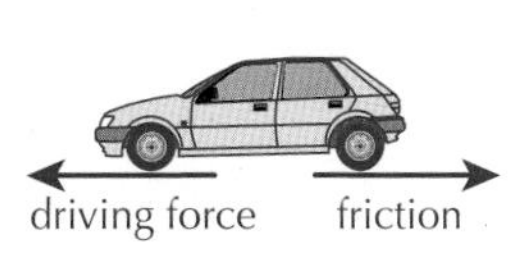

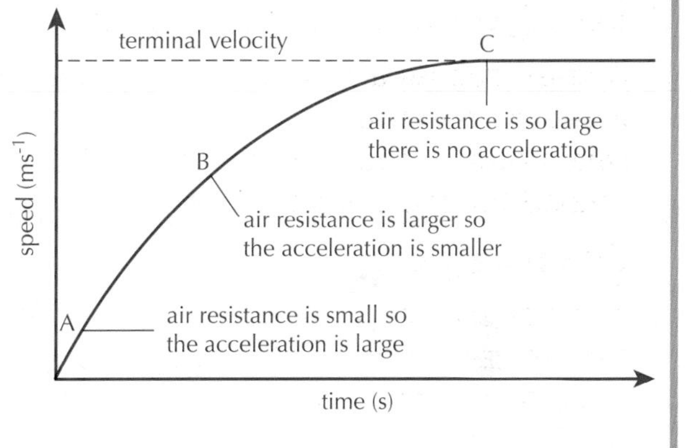

The final, constant speed of the car is known as terminal velocity.

Quick Test 12

1. Describe the effect of balanced forces on the motion of an object.
2. What does the size of the acceleration of an object depend on?
3. Why does a car have a top speed?
4. If a 2 kg block is pulled with a force of 10 N and friction is 2 N:
   (a) calculate the size of the unbalanced force
   (b) calculate the block's acceleration.

Matter and energy

Physics to learn	How to consider the terms matter, mass and energy.
Revision guide	You should be familiar with various forms of energy.

At the heart of physics

Physics involves the study of **matter** and **energy**. But what are these quantities? When we try to understand our universe, we gather information from matter and energy. Galaxies, stars, planets, humans, animals and molecules are all made of matter. These objects can interact by exchanging energy such as heat, light, sound and kinetic energy.

Energy cannot be created or destroyed, it can only be changed or transformed from one form to another. This is known as the **law of conservation**.

Matter and mass

Matter is usually the general name for the substance from which an object is made. Matter will have mass and volume.

Small particles – such as atoms – and sub-atomic particles – such as protons, neutrons and electrons – are the building blocks of matter.

Matter can exist in different states, e.g. solid, liquid, gas or plasma.

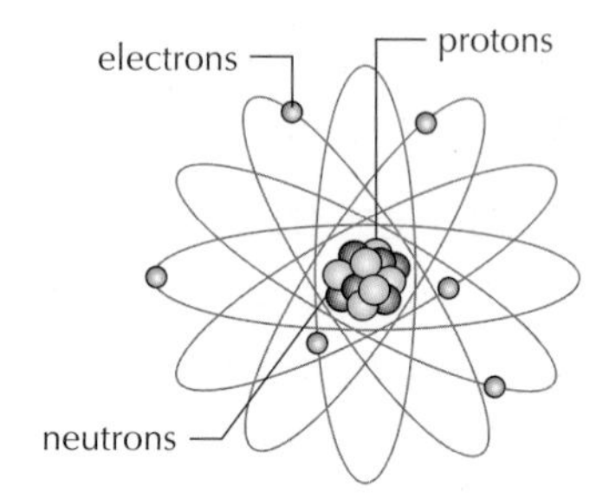

Mass defines the amount of matter in an object, measured in kilograms.

Energy

What are the different forms of energy?

Potential energy

This is the energy that is or can be stored in an object. This energy has the potential to do work e.g. a coiled spring can turn a clock, or the stretched rubber in a catapult can release a projectile.

- **Gravitational potential energy.** The energy an object has when it has been raised through the gravitational field. This energy will not just make an object fall towards the planet, but will make it accelerate, e.g. a skydiver jumping from a plane.

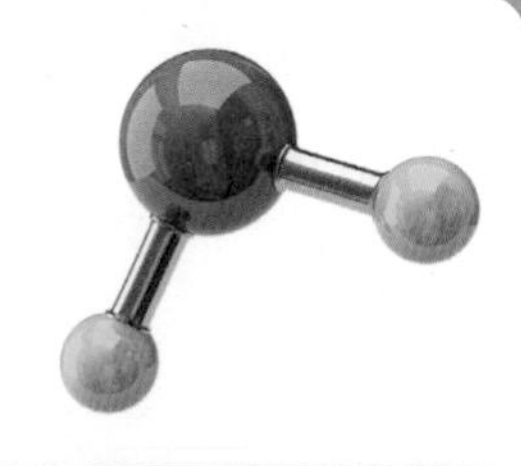

- **Chemical potential energy.** The energy stored within the bonds of atoms and molecules of a substance. A battery has chemical potential energy.

Kinetic energy

The energy of moving objects is called kinetic energy.

A car and its passengers need a lot of energy to get moving. They have gained kinetic energy. The car would keep moving if there was no friction to overcome during the journey. When brakes are applied the kinetic energy changes into a different form of energy: mainly heat energy and a little sound energy.

Heat energy

When something heats up, it is said to have gained heat energy. All atoms in any substance vibrate slightly, providing they are above absolute zero (0 kelvin). Increasing a substance's heat energy causes the atoms in that substance to vibrate faster.

Sound energy

Sound energy also occurs when atoms or molecules in a substance vibrate. The energy is passed from molecule to molecule, travelling through the substance (gas, liquid or solid) as a wave. A tuning fork or loudspeaker sends vibrations through the air. Sound energy cannot travel through a vacuum e.g. space, as sound needs a medium to travel through.

Light energy

Light is a wave of energy and is part of the electromagnetic spectrum. Light energy can travel through space.

Electrical energy

Electrical charges moving around a circuit are doing work. The charges move slowly but electrical energy passes through the charges at virtually the speed of light.

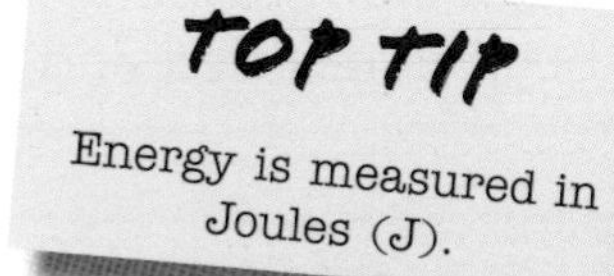

Nuclear energy

This is a type of stored energy that exists in the electrical bonds between particles of the nucleus in an atom.

Quick Test 13

1. Name **two** things that matter will have.
2. What is the unit of mass?
3. List **five** types of energy.
4. List **five** emitters of energy.
5. Research uses of energy.

Work done

Physics to learn	Work done, energy transferred.
Revision guide	You can describe work done and its effect on displacement.

Work and energy

Work is done when energy is transferred to an object or energy is transformed from one type to another. Work and all forms of energy are measured in joules, J.

Work done is equivalent to energy transferred.

Work, unbalanced force and displacement

A force is said to do work when it acts on a body so that there is a displacement of the body in the direction of the force.

A swimmer pushes off a board to accelerate forward. Push is a force and work is done while the force is applied.

A weightlifter raises a set of weights above his head. If the weights are 1000 N he will have to exert a force of 1000 N in the opposite direction. Lifting a displacement of 2 m does twice the work of lifting 1 m.

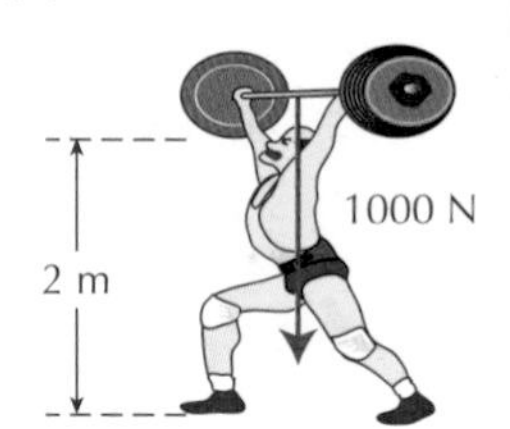

Here is an example of no work being done!

A person leans on and exerts a force on a wall. The wall also exerts a force on the person. As no displacement takes place, no energy has been transferred and so no work has been done.

Calculations

Work (E_W) is a scalar quantity, measured in joules (J).

We calculate the work done from the product of unbalanced force (F) and distance (d).

$$E_W = Fd$$

TOP TIP

Work is only done when a force is exerted over a distance.

Examples

1. A workman exerts a force of 2·3 kN on a wheelbarrow over a distance of 180 m.

$$E_W = Fd = 2300 \times 180$$
$$= 414\,000\,J$$
$$= 414\,kJ$$

2. A woman pushes her lawnmower a distance of 60 m and does 8 kJ of work.

$$E_W = Fd \quad 8000 = F \times 60 \quad F = 133\,N$$

3. When a car accelerates from rest to a high speed, the engine does work by exerting a large force during the distance travelled while accelerating. The work done changes to kinetic energy.

$E_W \rightarrow E_k$ $\quad Fd = \frac{1}{2}mv^2$ assuming no work is done against friction.

In real life, friction is transferring energy to the surroundings. Some work is done against friction, and the rest becomes kinetic energy.

$$F_{engine}d = F_{friction}d + \frac{1}{2}mv^2$$

4. A girl lifts a vase from the floor onto a shelf a height of 2 m above the ground. The vase then slips and free-falls to the ground. What energy changes are involved?

During the lift work is done on the vase and the object gains potential energy.

$E_W \rightarrow E_P$ $\quad Fd = mgh$

During the fall the potential energy changes to kinetic energy.

$E_P \rightarrow E_K$ $\quad mgh = \frac{1}{2}mv^2$

Note: The Potential and kinetic energy topic next should be studied along with this page.

TOP TIP

Work done is an energy conversion. Work and energy are both measured in joules (J).

Quick Test 14

1. A girl pushes her bike for 300 m with a force of 70 N. Calculate how much work she does.
2. A pupil weighing 500 N climbs the school stairs using 2500 J of energy. Calculate the height of the stairs.
3. A bag of weight 60 N is lifted onto a table of height 0·8 m. Calculate the work done and the potential energy gained.
4. A 1 kg ball leaves the player's foot at 30 ms^{-1}. How much kinetic energy did the ball gain?

Potential and kinetic energy

Physics to learn	Identify energy changes, understand potential and kinetic energy equations.
Revision guide	Select energy equations to use where energy is transferred.

Potential energy

To lift an object up through the gravitational field of the planet we need to do work. We have to exert a force that is equal to the weight of the object for the distance equal to the height the object is raised.

This work is transferred to the object as gravitational potential energy.

Weight is a force downwards and its size depends on both mass and gravitational field strength:

$W = mg$

The equation for calculating gravitational potential energy is found as follows:

Potential energy gain = Work done

$E_p = E_w$

$E_p = Fd$ (force × distance)

$E_p = mg \times d$ (mass × gravity [the size of force needed] × distance [the change in height])

So:

Potential energy = mass × gravity × change in height

$E_p = mgh$

Where:

- E_p is measured in joules
- m is measured in kg
- g is measured in ms^{-2} ($g = 9{\cdot}8\ ms^{-2}$ on earth)
- h is measured in m

TOP TIP

$E_p = mgh$ can give an increase or decrease in potential energy.

TOP TIP

Always check the units in every part of a question. If a quantity is given to you in g, you'll need to convert it to kg. If a distance is provided in cm, you will need to convert it to m.

Example

A forklift truck has lifted a 450 kg box by 5 m up to a shelf.

The gain in potential energy,

$$E_p = mgh$$
$$= 450 \times 9{\cdot}8 \times 5$$
$$= 22\,050\,J$$

Whenever we raise a mass up through the gravitational field we do work. The mass stores this as a gain in potential energy.

TOP TIP

Energy is a scalar quantity with no direction.

Kinetic energy

Kinetic energy is the energy that moving objects have. They have gained this energy while accelerating. Work is being done by applying a force for a distance while accelerating the object.

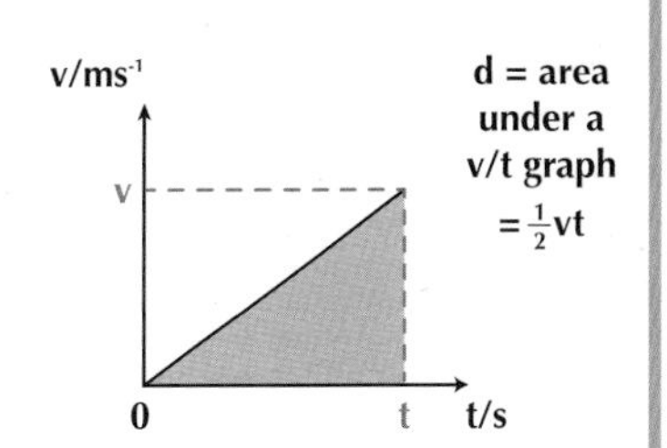

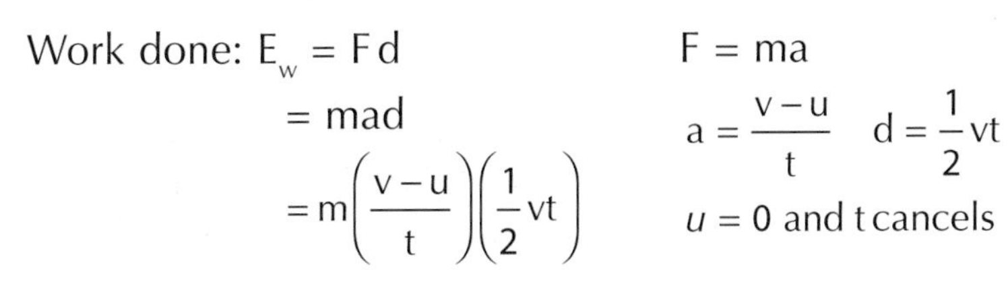

Work done: $E_w = Fd$

$= mad$

$= m\left(\frac{v-u}{t}\right)\left(\frac{1}{2}vt\right)$

$F = ma$

$a = \frac{v-u}{t} \quad d = \frac{1}{2}vt$

$u = 0$ and t cancels

so $E_k = \frac{1}{2}mv^2$

TOP TIP

Understand only the velocity is to be squared.

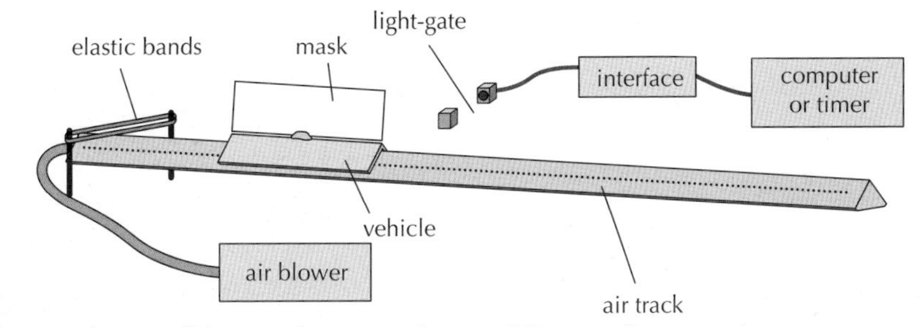

This air track vehicle is accelerated by the energy stored in the elastic bands. The vehicle gains kinetic energy.

The light gate is connected to a computer, which measures the **velocity**.

1. If the vehicle's mass is doubled, twice the energy (double the number of bands) is needed to get the same velocity. The experiment shows:

   Kinetic energy varies proportionately with mass: $E_K \alpha m$

2. To get the vehicle to double its velocity requires four times the energy (number of elastics). The experiment shows:

   Kinetic energy varies proportionately to velocity squared $E_K \alpha v^2$

Example

How much energy will stop a 1000 kg car going at 30 ms⁻¹ (70 mph).

$E_K = \frac{1}{2}mv^2 = \frac{1}{2}\,1000\,(30^2) = 500 \times 900 = 450\,000\,J.$

Quick Test 15

1. Calculate the work done to push a sledge a distance of 5m when 50N of force is required.
2. Calculate the gain in potential energy when a 3 kg bag is lifted onto a 2 m high shelf.
3. How much energy has a 5kg mass gained when it is moving at $2ms^{-1}$?

Energy conservation

Physics to learn	Equating energy equations.
Revision guide	Solve problems involving conservation of energy.

Conservation of energy

Energy is never created or destroyed – just changed from one form to another. We can use this principle, known as the conservation of energy, to predict the velocity reached by the cars at the bottom of this rollercoaster.

Assume:

$E_{P\ loss} = E_{K\ gain}$ $mgh = \frac{1}{2}mv^2$

Example

The cars, including their passengers, have a mass of 3000kg. If the cars descend a height of 50m, calculate the speed of the cars gained by this drop.

We assume that the kinetic energy gained is equal to the potential energy lost. When we equate these equations we see that mass appears on both sides. This means that the mass cancels out and so the mass does not actually matter.

$E_{p\ loss} = E_{k\ gain}$

$mgh = \frac{1}{2}mv^2$

$gh = \frac{1}{2}v^2$

$9{\cdot}8 \times 50 = \frac{1}{2} \times v^2$

$980 = v^2$

$v = 31{\cdot}3\ ms^{-1}$

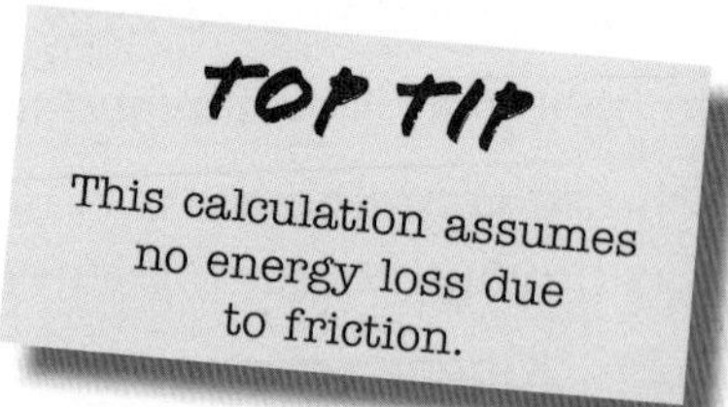

Energy and power

Power is the rate of doing work or the work done in unit time (1 s).

Power is the rate of transferring energy or the energy transferred in unit time (1 s).

$\text{power} = \frac{\text{work done}}{\text{time}}$ $\text{power} = \frac{\text{energy}}{\text{time}}$ $P = \frac{E}{t}$

Power is measured in watts (W), $1\,W = 1\,J\,s^{-1}$

Re-arranging this equation can give the energy transferred: $E = Pt$ or the time: $t = \frac{E}{P}$

Energy transfers

A crane lifts a load of bricks of mass 1200 kg onto a building of height 12 m. The carrier itself has a mass of 300 kg.

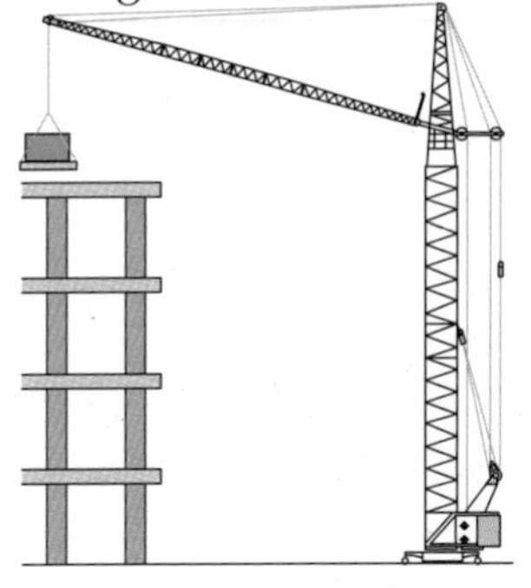

What minimum power must the motor of the lift develop to lift the bricks in 15 s?

Work done = potential energy gained ($g = 9{\cdot}8\ \mathrm{N\,kg^{-1}}$)

$E_p = mgh = 1500 \times 9{\cdot}8 \times 12 = 176\,400\ \mathrm{J}$

Power, $P = \frac{E}{t} = \frac{176\,400}{15} = 11760\ \mathrm{W}$

Energy 'loss', the great escape!

When a moving vehicle is stopped, the kinetic energy of the vehicle is mainly transformed to heat by the brakes. This heat energy will escape to the surroundings. Any energy that escapes as heat is hard to recover. It is usual to say the energy is 'lost'.

A skier is at the top of a 600 m high hill. She skis to the bottom, where she reaches a speed of $15\ \mathrm{ms^{-1}}$. If her mass was 66 kg, how much energy was lost because of **friction** during the descent?

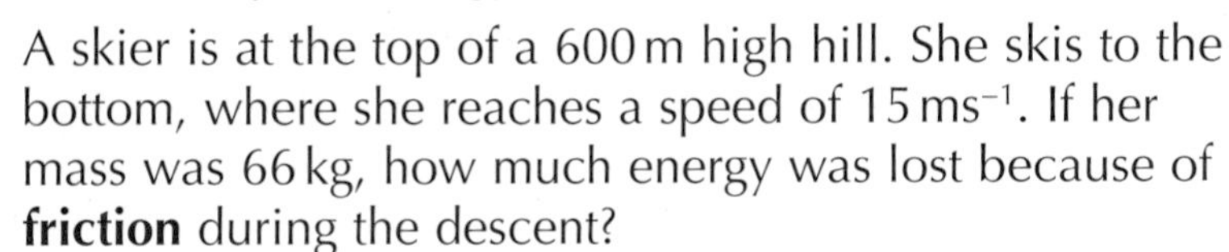

Her **potential energy** lost: $E_p = mgh = 66 \times 9{\cdot}8 \times 600 = 388\ 080\ \mathrm{J}$.

Her **kinetic energy** gained: $E_k = \frac{1}{2}mv^2 = \frac{1}{2} \times 66 \times 15^2 = 7425\ \mathrm{J}$.

We can see that this skier has used the force of friction to prevent her accelerating to too high a speed. Of her potential energy, 380 655 J has transferred to heat, and possibly a little sound.

From **conservation of energy:**
potential energy → kinetic energy + heat energy + sound energy.

Quick Test 16

1. What is the basic unit of time?
2. A dog pulls a sledge for 1500 m using a force of 50 N. Calculate the work done and calculate the dog's average power if it takes 10 minutes to pull the sledge that distance.
3. A 2 kW motor pulls a load of 3000 N 8 m. How long does it take?
4. A 2kg mass is dropped from a height of 3m. Calculate the speed at which it will contact the floor.

Mass and weight

Physics to learn	Mass and weight.
Revision guide	You can distinguish between mass and weight and identify gravitational field strength.

Mass

Mass is the amount of matter there is in an object. Mass depends not only on the size of an object but on what it is made from. (If it has a high density, it will have more mass.)

On the atomic scale, mass depends on the number and type of atoms.

Mass is scalar – it has only magnitude (size). We measure mass in kilograms (kg).

Gravitational field strength

An invisible field is said to exist around every mass, which we call the gravitational field. We say the field exists because we can see the effect of placing an object near a large mass such as the Earth. An object will accelerate towards the large mass. A mass requires a force to accelerate.

On Earth, the weight of every 1 kg of mass is 9·8 N. We can use 1 kg of mass as a fair test method of checking the field strength. The force (weight) on 1 kg of mass is measured. This weight per unit mass is equal to the gravitational field strength (g) and has the units $N\,kg^{-1}$.

The acceleration caused by gravity is numerically equal to the gravitational field strength.

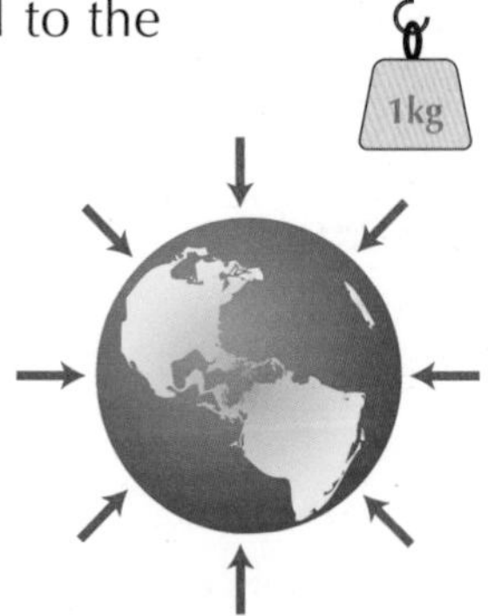

On Earth the following apply: acceleration, $a = 9{\cdot}8\,ms^{-2}$ and gravitational field strength, $g = 9{\cdot}8\,N\,kg^{-1}$.

The gravitational field strength decreases as you move away from the surface of a planet or star.

For the first few kilometres from the Earth we do not notice the field strength drop much as the Earth is so much larger, but a spacecraft or satellite will observe the decrease in strength. Gravitational field strength is the force on unit mass. $g = \frac{F}{m}$

On Earth, 5 kg will weigh 49 N, $g = \frac{F}{m} = \frac{49}{5} = 9{\cdot}8\ N\,kg^{-1}$, i.e. 9·8 N on each kg.

TOP TIP

The larger the mass, the larger the strength of its gravitational field.

TOP TIP

You can look up the gravitational field strengths you need in the exam paper.

Weight

Weight is a force.

Weight acts downwards. Weight is a vector. Weight depends on both mass and the gravitational field strength. $W = mg$

Examples

A pile of rocks has been collected from the Moon. They contain 4 kg of matter.

On the Moon their mass is 4 kg and their weight is:

$W = mg = 4 \times 1{\cdot}6 = 6{\cdot}4\,\text{N}$.

Back on Earth, the rocks have the same 4 kg mass, but their weight has increased:

$W = mg = 4 \times 9{\cdot}8 = 39{\cdot}2\,\text{N}$.

TOP TIP

The mass does not change but the weight changes with gravitational field strength.

Planets

Objects in our solar system	Gravitational field strength on the surface (N kg^{-1})
Earth	9·8
Jupiter	23
Mars	3·7
Mercury	3·7
Moon	1·6
Neptune	11
Saturn	9·0
Sun	270
Uranus	8·7
Venus	8·9

Example

An object weighs 150 N on Mars.

What is its mass?

What would it weigh on the surface of Jupiter?

$W = mg \quad 150 = m \times 3{\cdot}7$

mass of object, $m = 40{\cdot}5\,\text{kg}$

$W = mg = 40{\cdot}5 \times 23 = 932{\cdot}4\,\text{N}$

Quick Test 17

1. When g changes, does mass or weight change?
2. Calculate the weight of the following masses on Earth:
   (a) 750 kg **(b)** 1×10^3 kg **(c)** 450 g
3. Calculate the mass of the following weights on the Moon:
   (a) 32 N **(b)** 9500 N **(c)** 3 kN
4. Calculate the weight of 1000 kg on Mercury.

Projectile motion

Physics to learn	Projectile motion and applications.
Revision guide	You can explain and do calculations on projectile motion.

Projectiles from space

Projectile motion occurs once an object has been put into motion by a force and then the only force acting on it is the force of gravity. The path of the projectile is known as its **trajectory**.

Consider a space vehicle before it re-enters our atmosphere.

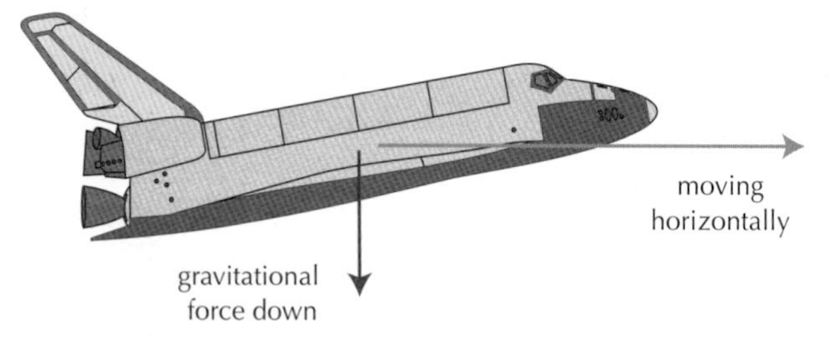

TOP TIP

The horizontal motion is independent of the vertical motion.

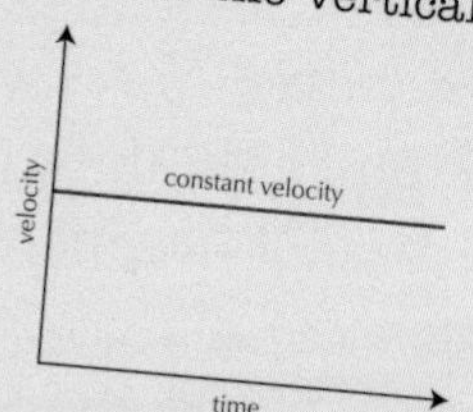

For the constant horizontal velocity:

$$v_h = \frac{s}{t}$$

area under v_h-t graph gives horizontal range

Assume no friction acting.

The space vehicle is travelling horizontally at a constant velocity.

The force of gravity pulls on the space vehicle in a vertically downwards direction. The force of gravity creates acceleration downwards.

The resultant velocity at any point in the trajectory is a combination of both the constant horizontal velocity and the increasing vertical velocity at that point.

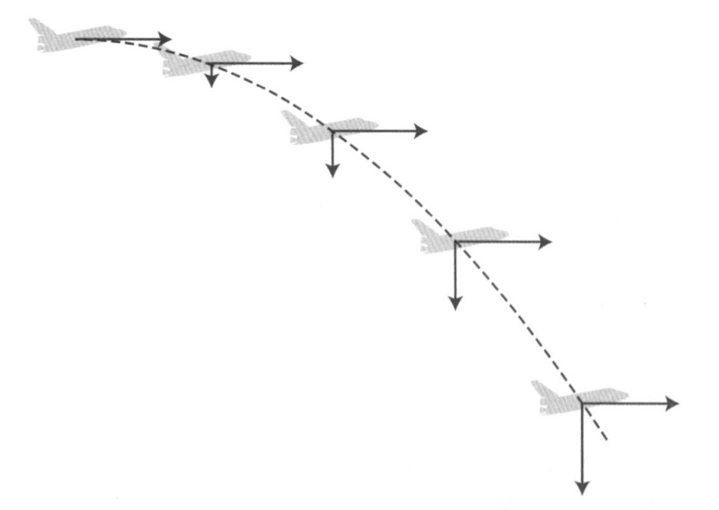

The space vehicle follows a projectile trajectory.

TOP TIP

The vertical motion is independent of the horizontal motion.

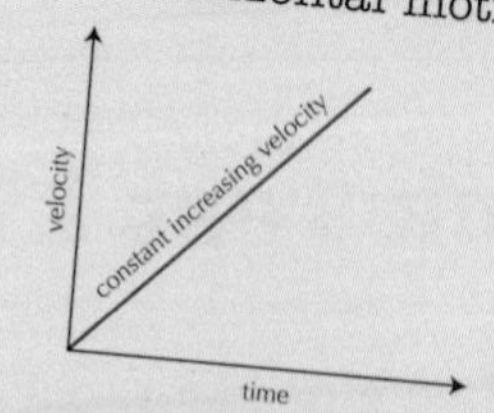

Constant vertical acceleration gives increasing vertical velocity:

$$v_v = u + at$$

area under v_h-t graph gives vertical height

Projectiles fired horizontally

What is important to remember is that the motion along the horizontal direction does not affect the motion along the vertical direction and vice versa. Horizontal motion and vertical motion are totally independent of each other.

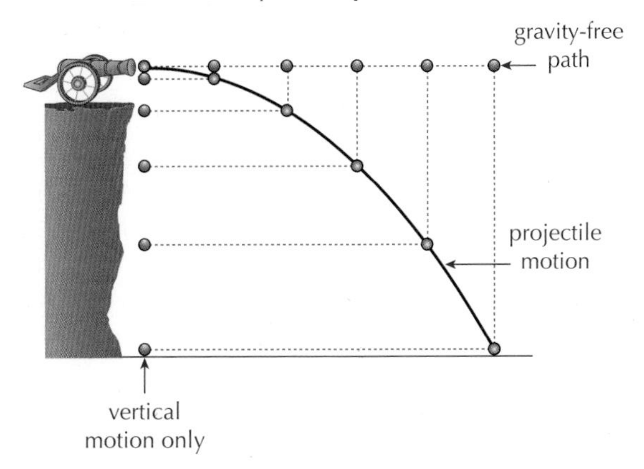

- The horizontal motion remains constant velocity.
- The vertical motion is constant acceleration.
- Note both the falling and the projected objects will hit the ground at the same time.

Example

A ball is falling off a cliff with a velocity of $7 ms^{-1}$. It is in flight for 3s.

What range does it have?
What is the resultant velocity on impact?

Horizontally:

$V_H = 7 ms^{-1}$ $\quad t = 3s$

$d_H = vt = 7 \times 3 = 21m$

Range = 21m

Vertically:

$uv = 0 ms^{-1}$ $\quad t = 3s$ $\quad g = -9{\cdot}8 ms^{-2}$

$v = u + at = 0 + (-9{\cdot}8)3 = 29{\cdot}4 ms^{-1}$

The resultant velocity from vector addition is $30{\cdot}2\ ms^{-1}$ at $76{\cdot}6°$

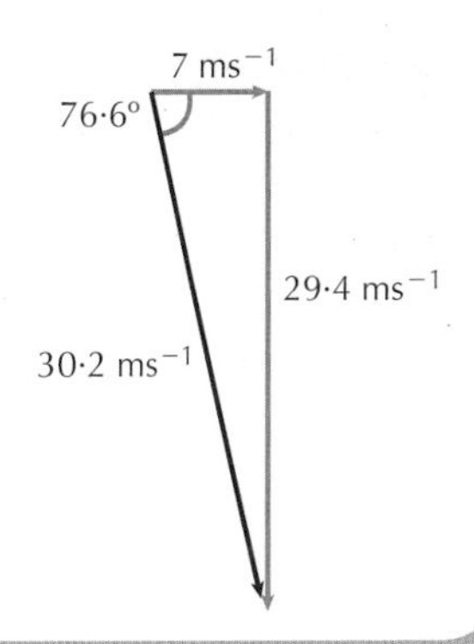

Quick Test 18

1. State what forces are acting on a projectile.
2. State the direction of that force.
3. What is the direction of the acceleration?

Satellites

Physics to learn	Newton's thought experiment and satellites.
Revision guide	You can explain satellite orbits.

Satellites

A satellite is usually thought of as a smaller body orbiting a larger body. Satellites can be divided into two groups: natural satellites and artificial satellites.

Natural satellites

Moons are natural satellites. A natural satellite orbits a planet or other celestial object.

Artificial satellites

An object that has been put into space by human effort is an artificial satellite.

The first artificial artificial ever launched was the Sputnik 1 from the Soviet Union in 1957. Since then, over 8000 artificial satellites have been launched, although most of these are now space debris.

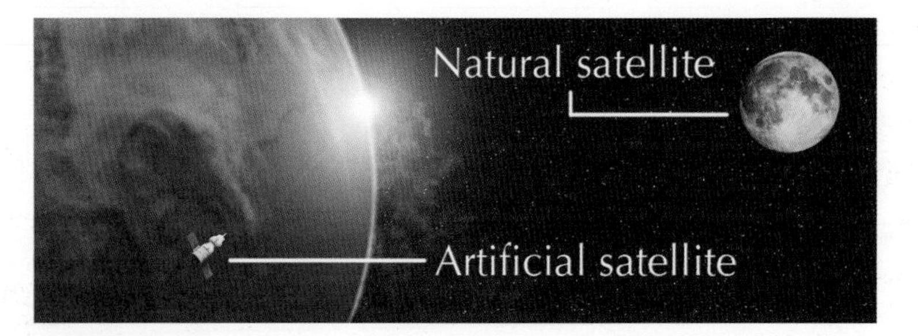

Newton's thought experiment

Why does the apple fall to the ground but the Moon stay in orbit?

Isaac Newton used the idea of a cannonball being fired from the top of a very high mountain and parallel to the Earth.

If there was no gravity, then the cannonball should follow a straight line away from Earth.

However, Newton knew that a cannonball would normally follow a projectile trajectory. If the initial velocity was increased, Newton illustrated the increased path that the cannonball would take. A projectile with a fast enough horizontal velocity could fall to earth at the same rate as the earth curves away.

Satellite motion extends the ideas of projectile motion.

Satellite motion was first predicted by Newton in his **thought experiment** over 300 years ago.

Newton thought if a cannonball was fired off a very high mountain fast enough it would never reach the ground. Instead it would remain in free-fall towards the Earth but, because the Earth is round, the cannonball would remain in orbit as an artificial satellite.

The apple falls to the ground and satellites remain in orbit because of the same gravitational force.

TOP TIP

Research Newton's thoughts on gravity and the planets.

Satellite orbits

Today satellites are launched into orbit in space at great heights. Earth's mountains are not high enough!

The highest mountain on Earth is Mount Everest, at 8848 m, which is nearly 9 km.

Orbits for useful satellites are much higher than this. TV satellites in geo-stationary orbit are found at 36 000 km.

Low Earth orbit is up to 2000 km.

Medium Earth orbit is from 2000 to 36 000 km.

High Earth orbit is above 36 000 km.

TOP TIP

The Earth has a radius of about 6000 km.

A geostationary satellite has a period of 24 hours and orbits at a height of 36 000 km. Communications and weather forecasting satellites orbit the earth at this height so they stay above the same area of earth.

The period of a satellite in high altitude orbit is greater than the period of a satellite in low altitude orbit.

Other benefits of satellites include GPS, scientific discovery and telescopes for space exploration.

Quick Test 19

1. What sensation do you experience in free-fall?
2. In which direction does a projectile have:
   (a) constant speed?
   (b) acceleration?
3. How can a satellite be projected into orbit?
4. Write out a definition of a natural and an artificial satellite.
5. At what height are the two boundaries between the different orbit bands?

Space exploration

Physics to learn	Current understanding of our universe.
Revision guide	Can follow developments in space exploration.

Space travel

Challenges

Adjustments to speed and direction in space require a force to be exerted.

Ion propulsion

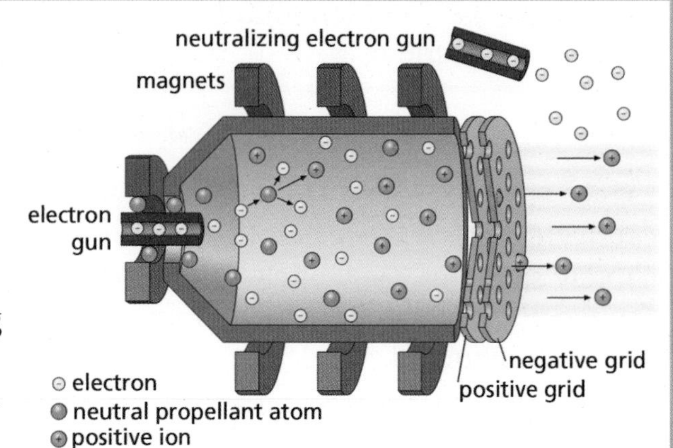

Ion propulsion systems have been in ongoing development since the late 1950's. They create thrust by accelerating ions or plasma with electricity.

Electric propulsion allows a high velocity to be achieved by producing a small unbalanced force over a long period of time. They are useful in the near vacuum of space but a chemical burn is still required to escape the Earth's atmosphere.

Charged ions are accelerated out of the thruster as an ion beam, which produces thrust. (There is no overall build up of charge so the spacecraft remains neutral.)

Ion thrusters have enabled spacecraft to travel deep into our solar system. Around the turn of the millennium, the first spacecraft to travel mainly by ion propulsion, travelled 163 million miles on the "Deep Sea 1" mission.

Ion thrusters are also being used to keep satellites in their desired locations.

Continuing advancements will adapt ion thrusters for a broad range of missions.

Gravitational catapults

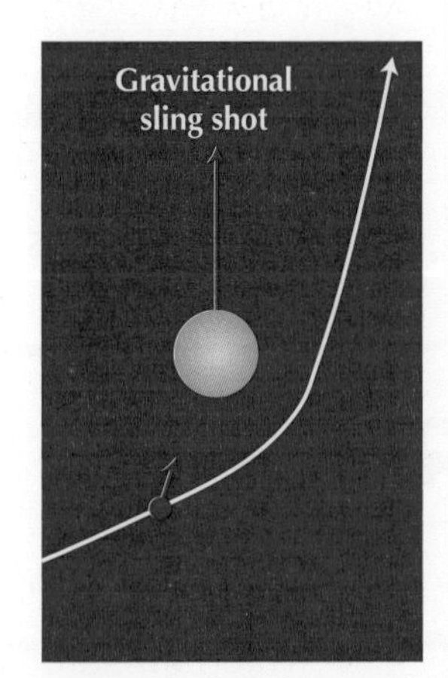

If a satellite or spacecraft approaches an asteroid or planet that is in motion, it will gain a sling shot from its orbital velocity.

Risks

Re-entry risks

Materials on a spacecraft cannot be allowed to melt or change state on re-entry to the earth's atmosphere. Special silica tiles may be used to absorb the heat.

Space supplies

Space craft making manoeuvres to the International Space Station with a large mass of thousands of kilograms creates potential risk.

TOP TIP

Now list some challenges and risks for space travel.

Newton's laws

Newton's laws apply throughout space travel.

Newton's third law

If the rocket exerts a force on the fuel, the fuel exerts a force on the rocket.

Newton's second law

The fuel pushing on the rocket causes acceleration.

Newton's first law

No force is required for an object to continue to travel at constant velocity.

Gravitational field strength

Gravitational field strength is defined as the force on unit mass. $g = \frac{f}{m}$. On earth g = 9·8 N m-1.

Gravitational field strength decreases with height as we move away from a planet.

TOP TIP

Look up the values for g for various objects under "Mass and weight". To find our weight, our force on a planet, use: $W = mg$

The observable Universe

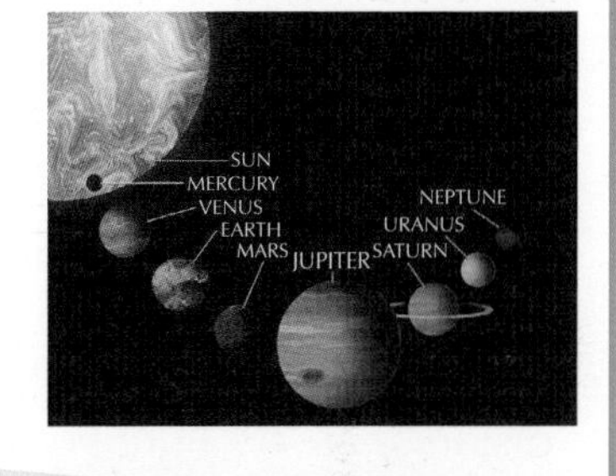

A Solar System consists of a star and, in orbit around it, planets, moons, asteroids and comets. Our star, the Sun, contains over 99% of the mass of our Solar System.

Earth is one of eight planets: Mercury, Venus, Earth, Mars, Jupiter, Saturn, Uranus and Neptune. Four of these planets are small and rocky, then there is the asteroid belt, then four giant planets made of gas. Pluto is a dwarf planet.

Here is a memory aid for the order of the planets:

My **V**ery **E**xcellent **M**um **J**ust **S**erved **U**s **N**achos.

Our Solar System is a tiny part of the Universe.

TOP TIP

In the Big Bang theory, red shift and cosmic microwave background radiation suggests that the Universe is expanding and has a beginning.

Planet	A massive body orbiting a star, made round by its own gravity.
Dwarf planet	A body which has not cleared other material around its orbit.
Moon	A natural satellite of a planet.
Asteroid	A small, rocky object that orbits the sun.
Sun	A star (one of 100 000 million in our galaxy!)
Star	Emits light and heat radiation.
Solar System	The Sun and the objects that orbit it.
Exoplanet	A planet orbiting a star other than our own Sun.
Galaxies	Consist of millions of stars.
Universe	Contains billions of galaxies.

Cosmology

Physics to learn	Light years, the universe and the big bang theory.
Revision guide	Convert between light years and metres. Describe the big bang theory.

Light years

Distances are so vast in space that astronomers use a unit for distance called the light year.

1 light year = the distance travelled by light in 1 year.

You can calculate how far this distance is in metres using $d = vt$.

$d = vt = (3 \times 10^8) \times (1 \times 365 \times 24 \times 60 \times 60)$
$= 9{\cdot}46 \times 10^{15}$ m

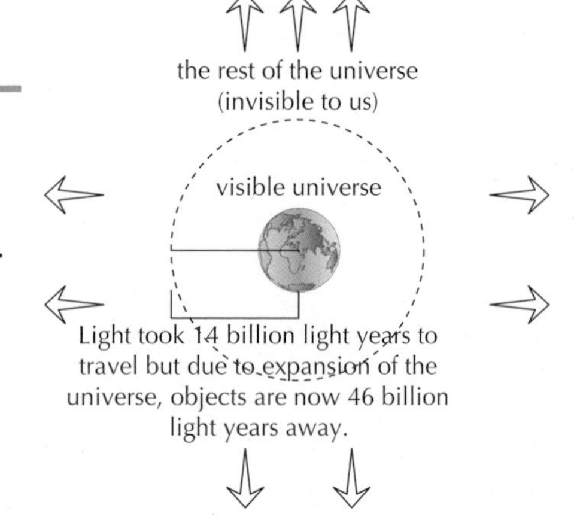

Here are some approximate distances:

Earth to Moon	1·2 light seconds
Earth to Sun	500 light seconds (8 minutes)
Our Sun to the next star	4·3 light years
Across our galaxy	100 000 light years

The closest star to our Sun is 24 million million miles away. That is 38 000 000 000 000 km. To measure the distance as 4·3 light years is more practical.

When our Sun goes below the horizon, it actually did so 8 minutes ago!

Modern space probes and the Planck telescope have provided data which has been used to calculate the age of the Universe – about $13{\cdot}8 \times 10^9$ years or 13·8 billion years.

TOP TIP

A light year sounds like a time but we must remember this is a distance!

TOP TIP

Time must be in seconds to work out distance in metres for a light year.

Telescopes for the Universe

Information about astronomical objects comes from observing visible light and also from space telescopes and observatories, which are used to detect radiations from the different parts of the electromagnetic spectrum.

Gamma ray

The Fermi Gamma-ray Space Telescope detects shortwave gamma rays, which are emitted by neutron stars, pulsars and black holes.

X-ray

The Chandra X-ray Observatory detects even faint X-rays.

Ultraviolet, visible, infrared

The Hubble space telescope carries telescopes for detecting ultraviolet, visible and infrared radiations. Spitzer is an older infrared telescope.

Microwave

The Planck telescope plots microwave background radiation.

Radio

Radio telescopes can be used singly or in an array.

TOP TIP

Research the uses of different telescopes.

TOP TIP

Research the uses of space technologies: Goretex, memory foam, zero gravity.

The big bang theory

Observations of galaxies show that they are moving away from each other. Reverse this discovery and the big bang theory says that the universe had a beginning with all matter and energy contained in a singularity, a tiny point like hot dense state, which then followed periods of expansion.

A timeline of the universe based on the Big Bang theory

Some key points to the theory:

- There was rapid expansion till sub-atomic particles were formed
- 400 000 years later these formed into atoms
- A further period existed, known as the Dark Ages, with no light
- Atoms combined, first stars formed, and fusion produced light from the stars
- Black holes, stars and galaxies followed
- The universe continues to expand

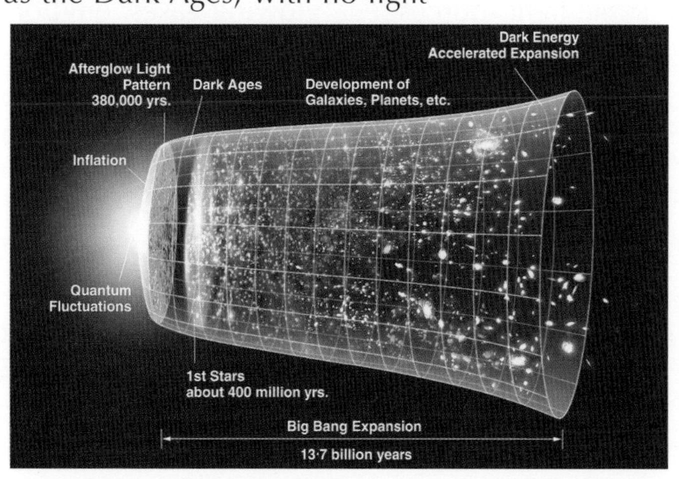

The latest big bang theory provides a calculation that the age of our universe is 13·8 billion years old.

Spectra

Physics to learn	You can identify continuous and line spectra.
Revision guide	You can distinguish how continuous and line spectra are produced and can identify elements in the stars.

Spectra

White light can be split (dispersed) into its different colours using a prism. The prism refracts short waves most.

Violet has the shortest wavelength, 400 nm.

Red has the longest wavelength, 700 nm.

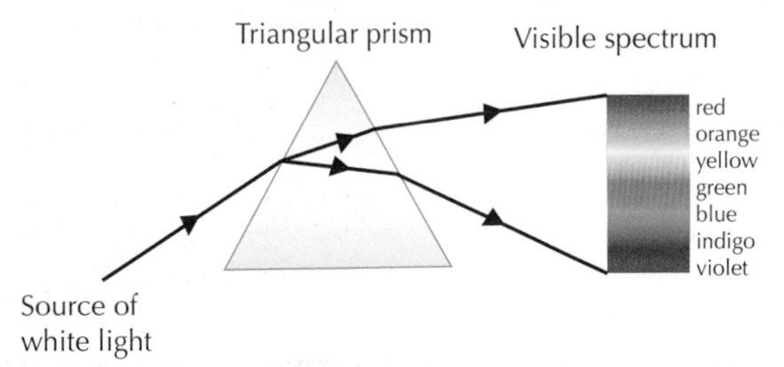

Spectroscopy

A spectroscope helps us find out what stars are made of. It disperses, or separates, light from a star into a very wide spectrum of the colours which make up the light it emits.

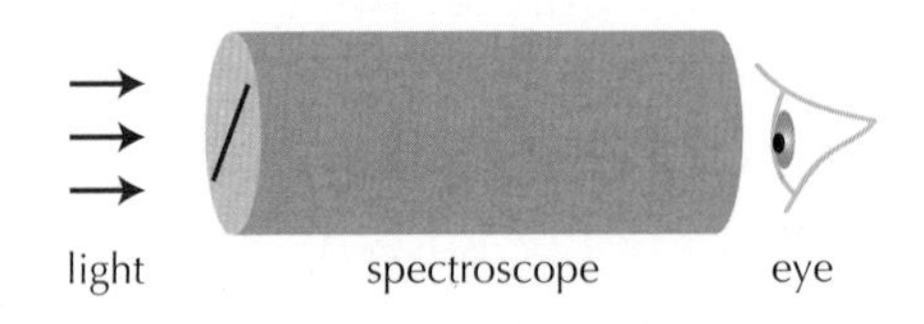

Continuous spectra

The Sun and filament bulbs emit a continuous spectrum containing all the wavelengths of visible light.

Line spectra

Some light sources, e.g. glowing gases, emit only certain wavelengths. They produce a line spectrum.

Light comes from within the structure of the atom. Each type of atom has its own set of electrons in orbit. Electrons have energy, which they can emit as light. Each atom has its own structure and unique line spectrum pattern.

There are two types of line spectra, emission or absorption. A line emission spectra shows colour lines over a black background. A line absorption spectra shows black lines over a continuous spectra. Either line emission and line absorption spectra will identify the elements present.

Understanding the Universe

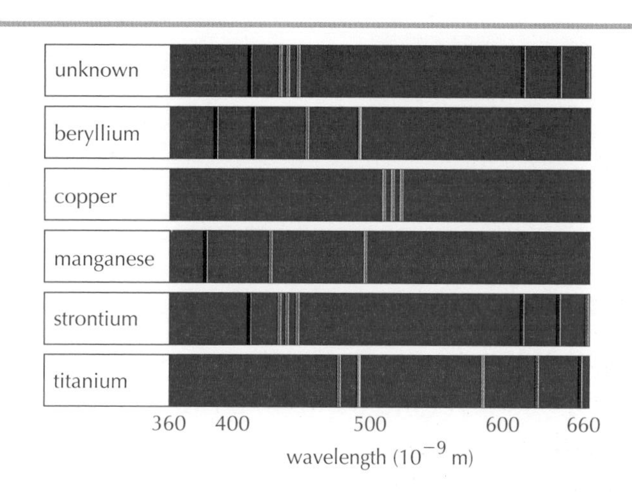

Astronomers look at the line spectra of light to identify elements present in distant stars.

From this chart you should be able to identify which element the unknown spectra is from. The unknown spectra comes from strontium.

Sometimes the spectra of a star appears to have shifted. This can tell us whether the star is receding or approaching us.

The diagram below shows an absorption spectrum, which contains black lines over the visible spectrum. The diagram also shows two examples of red shifted spectra, note the black lines have the same pattern.

> **TOP TIP**
> Albert Einstein made his discoveries when he was wondering about the nature of light.

We can learn a great deal about the gases and elements that exist in the Universe from spectroscopy. We should also stop and wonder about the brilliant relationship that is revealed between matter and energy. How is it that matter can emit light? This is a question that you will pursue in further courses in physics.

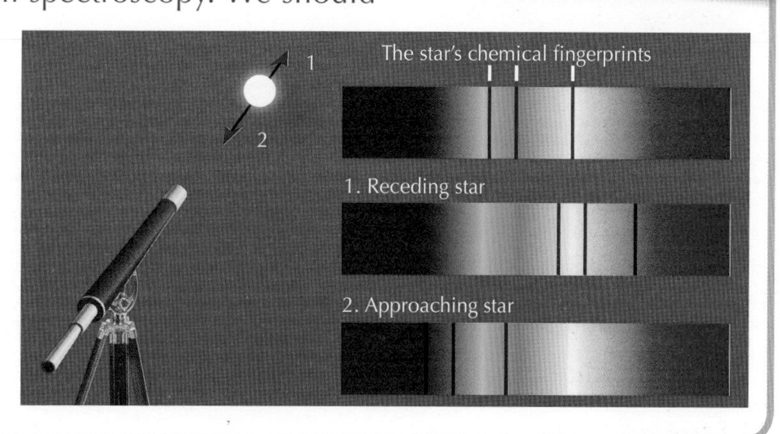

Quick Test 20

1. Why do atoms have their own individual spectra?
2. What type of spectra can be used to identify the elements?
3. What is the order of the colours of the continuous spectra?
4. Elements have only one type of __________.

Electrical charge and current

Physics to learn	Electric charge and electric current.
Revision guide	Understand charge and how to define current.

Charges

TOP TIP

Protons and neutrons cannot normally leave an atom. Electrons are much smaller and can be easily stripped from the outermost shell of electrons in an atom.

The atom:

In a neutral atom, the number of protons is equal to the number of electrons and so the atom has no overall charge. When electrons either leave or are added to an atom, this balance is upset. When an atom **gains** electrons it becomes negatively charged. When it **loses** electrons it becomes **positively** charged.

An atom has a nucleus surrounded by shells of electrons.

The **electrons** are found in shells around the **nucleus**.

The **nucleus** is found at the centre of the **atom** and contains **neutrons** and **protons**.

Conductors and insulators

In a conductor, electrons are free to move.

In an insulator, electrons are not free to move as they are all bound in the atom.

Good conductors are metals (e.g. gold, silver, copper and aluminium), and carbon.

Insulators are usually non-metal, e.g. pvc, polythene, wood, rubber and paper.

Electric charge

The electron is extremely small. Electrons carry a negative charge.

Electric charge is measured in coulombs (C).

The charge of an electron is $1{\cdot}6 \times 10^{-19}$ C.

In an electric circuit we do not measure the number of electrons flowing in the circuit but the quantity of charge.

The unit of charge

Charge is given the symbol Q and is measured in Coulombs C.

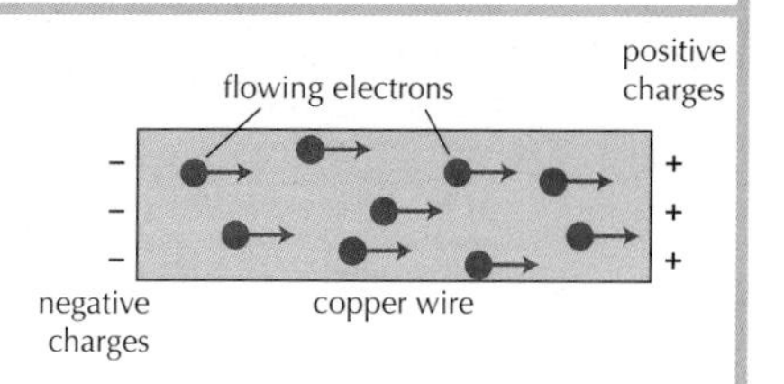

Electrons are tiny and have a very small amount of charge.

Each Coulomb of charge will contain roughly 6×10^{18} electrons. You need not remember this number. You may want to think of each Coulomb of charge as representing the charge from a large bundle or amount of electrons.

Electric current

When charges, e.g. electrons, flow we have an electric current.

Current is the rate at which charge flows and is the amount of charge transferred in unit time (1 s).

$$I = \frac{Q}{t} \qquad Q = It$$

Current (I) is measured in amperes (A).

Charge (Q) is measured in coulombs (C).

Time (t) is measured in seconds (s).

For example, 6 C passes a point in 2 minutes. What is the value of the current?

$$I = \frac{Q}{t} = \frac{6}{2 \times 60} = \frac{6}{120} = 0{\cdot}05\text{A}$$

TOP TIP

Current does not flow. Current is a measure of the quantity of charge flowing.

Measuring current

An ammeter is placed in line (in series) with the circuit.

I = 2 A. This means 2 C of charge flows through the bulb every 1 s.

How much charge flowed when there was a current of 15 μA in a wire for 3 hours?

$Q = It = 15 \times 10^{-6} \times (3 \times 60 \times 60) = 0{\cdot}162 = 0{\cdot}16\text{C}$

TOP TIP

Do not use amps or secs in your answers – they are not allowed. Instead, learn to use the symbols A and s.

Quick Test 21

1. What type of charge is on:

   (a) a proton **(b)** a neutron **(c)** an electron?

2. What type of charge flows in electric circuits?
3. Why are electrical wires made of copper and covered in plastic?
4. A circuit current is 0·5 A. Calculate how much charge passes in 3 minutes.

d.c. and a.c.

Physics to learn	How to describe differences in alternating and direct current.
Revision guide	You can give examples of d.c. and a.c. sources and distinguish d.c. and a.c. traces.

Electrical supply

There are two types of power supply: direct current (d.c.) and alternating current (a.c.).

d.c.
direct current

+ − cell

+ − battery

+ − power supply

For direct current, charges flow in one direction round a circuit.

Examples of direct current: anything that uses battery power is using d.c.

A cathode ray oscilloscope (CRO) shows what the voltage across a steady d.c. supply looks like:

d.c.

voltage 0

time

direct voltage => direct current

a.c.
alternating current

power supply

signal generator

For alternating current, charges repeatedly change direction round a circuit.

Examples of alternating current: anything that uses mains power is using a.c.

A cathode ray oscilloscope (CRO) shows what the voltage across an a.c. supply looks like:

a.c.

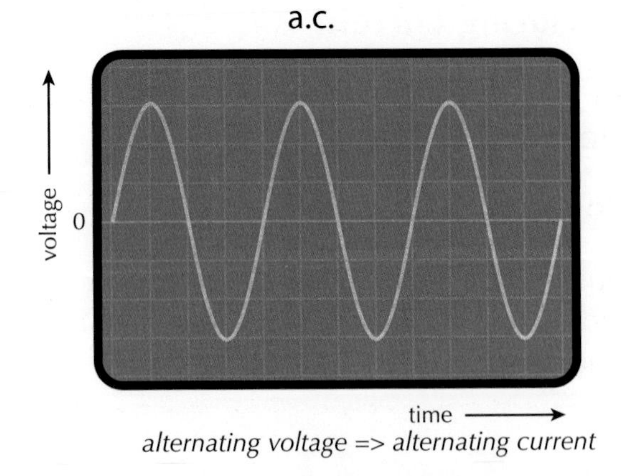

alternating voltage => alternating current

TOP TIP

The CRO measures voltage on the y-axis and time on the x-axis.

Mains voltage

The frequency of the mains is 50 Hz. There are 50 complete waves every second. The flow of electrons increases and decreases, then increases and decreases in the opposite direction, 50 times every second. (Each complete cycle lasts $\frac{1}{50}$ s.)

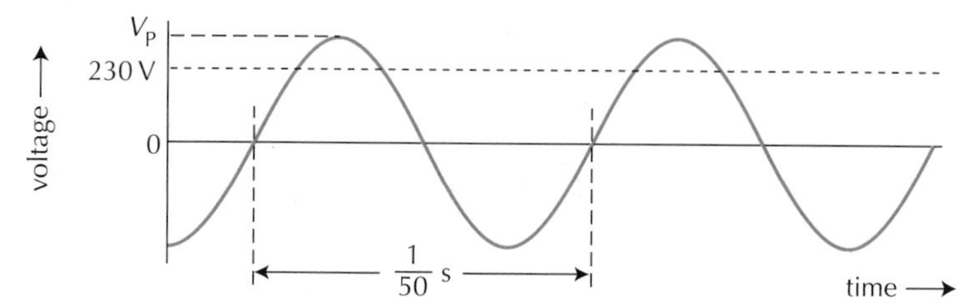

$f = 50Hz$

$t = \frac{1}{50} s$

$V = 230V$

$Vp = 325V$

For d.c., voltage and current are in one direction.

For a.c., the direction of voltage and current repeatedly reverses. The value of voltage and current varies between zero and a peak value. Peak value is always greater than the rated or quoted value. The quoted value is the effective value.

TOP TIP

The rated value of the mains (V = 230V) is its effective value.

Voltage and current

A dc current will have a dc voltage supply to push the charges round a circuit in one direction.

An ac current will have an ac voltage supply to push the charges round a circuit in alternating directions.

Most portable devices use a dc supply such as a battery.

Our homes are supplied with ac as there is a long distance between power stations and our homes. To use a small portable device in our homes we will normally be required to plug in an adapter.

Quick Test 22

1. What is the frequency of the Scottish mains supply?
2. If a battery is reversed, what happens to its signal on the CRO?

Electric fields

Physics to learn	Electric fields.
Revision guide	Know how a charged particle behaves in an electric field.

Electric fields

There is an electric field round electric charges.

Field lines are drawn away from a positive charge and towards a negative charge.

A charged particle experiences a force in an electric field.

The path of a free positive charge near single point charges.

The field lines show the direction a free positive charge will move. An electron, having a negative charge, will move in an opposite direction to these field lines.

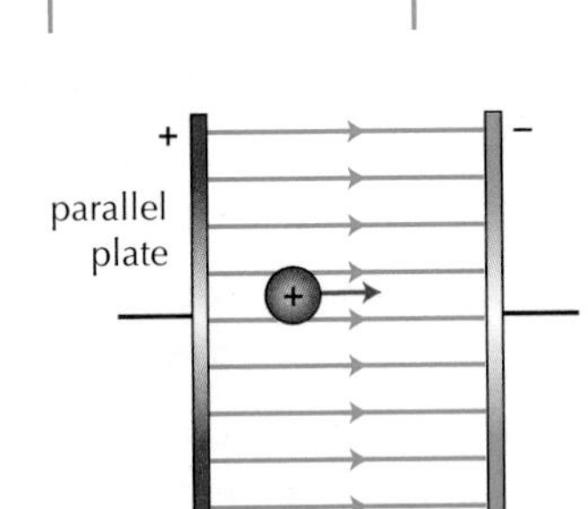

The path of a free positive charge in a uniform field.

Between two oppositely charged parallel plates there is a uniform field.

The field between two oppositely charged points:

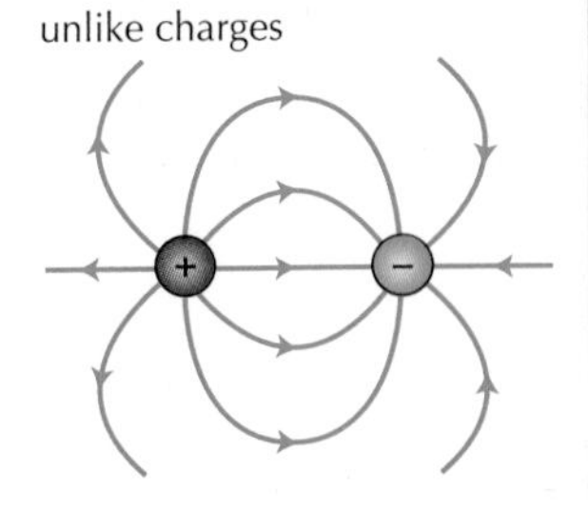

The field between two similarly charged points:

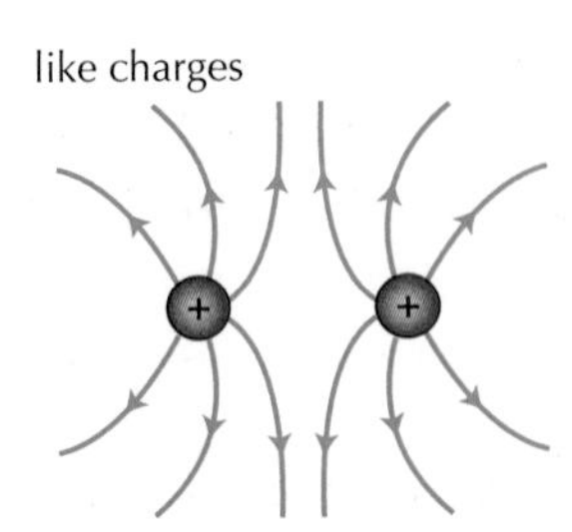

The force experienced by a free charge in a field will cause it to accelerate. An electron, having a negative charge, will accelerate in an opposite direction to these field lines. In all diagrams, a free positive charge will move with the arrows but an electron will move against them.

TOP TIP

The field between two negative charges would be similar but the arrows would be reversed.

Teltron tubes

Teltron tubes contain a vacuum which allows the path of free charges to be studied when in an electric field.

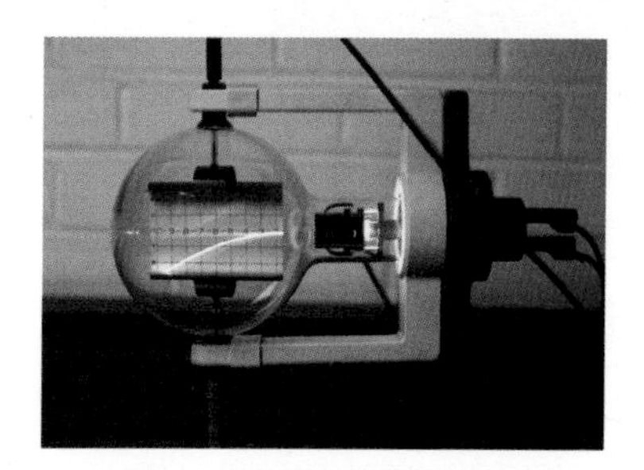

Here a beam of electrons enters a vertical field at right angles. The path of the electrons is shown on the screen. The electrons have been accelerated downwards. This combined with their original motion, produces the curved path seen.

Electrostatics

Rubbing two insulators together can electrically charge them by transferring electrons (negative charges) between them through friction. To make objects negatively or positively charged, electrons have to be added/removed.

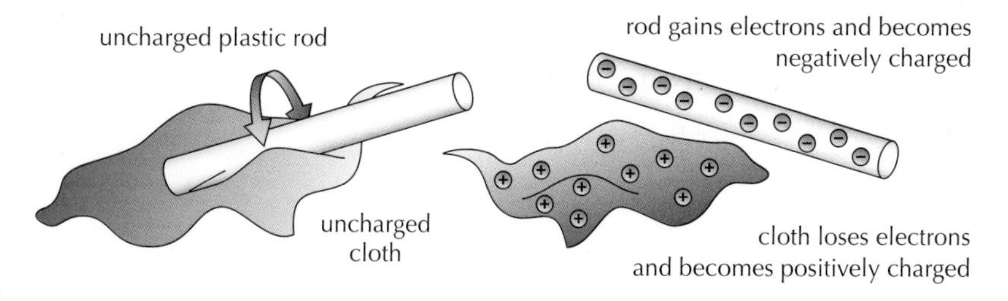

TOP TIP

Like charges **repel.**
Unlike charges **attract.**

Quick Test 23

1. What does charge experience in an electric field?
2. What does a force do to a charge?
3. What type of charge will move in a direction opposite to the field lines?
4. When a charge enters a field at right angles, what type of path does it follow?
5. A proton and electron are close together. Will they attract or repel?

Voltage

Physics to learn	Electric fields, voltage and potential difference.
Revision guide	Explain how charge experiences a force, describe and measure voltages.

Electric field

The region around an electric charge is an **electric field**.

In an electric field, an electric charge experiences a force.

Force causes acceleration, and work done by the field on charges creates kinetic energy in the charges. The field lines show the direction in which a positive charge will move, so in a circuit electrons flow against the field lines.

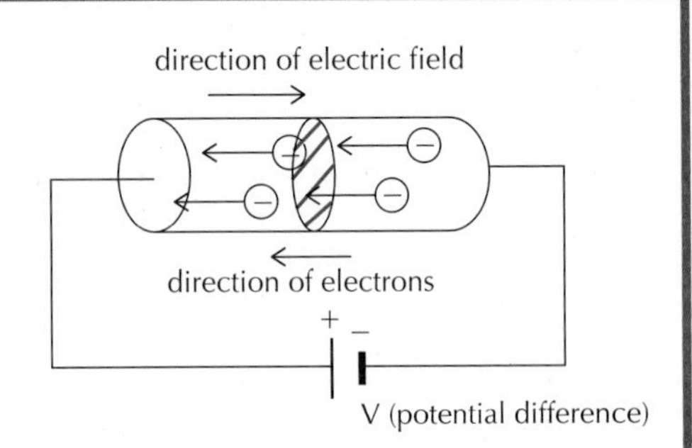

Circuit symbols

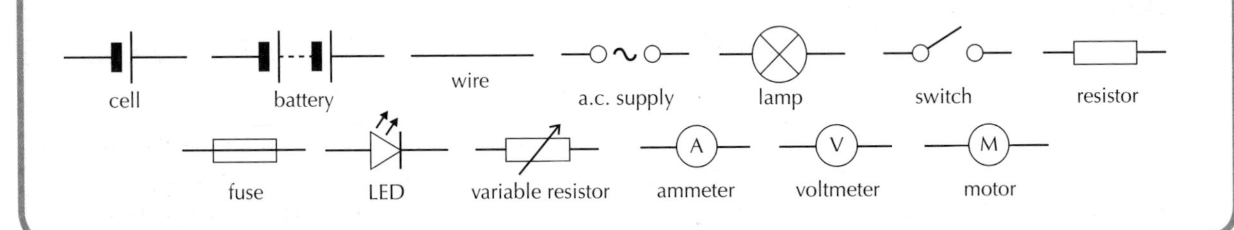

Voltage

The voltage of a supply is a measure of the energy given to the charges in a circuit.

The voltage (or potential difference, p.d.) across a component is a measure of the energy given out by charges as they go through the component.

Voltage (V) or p.d. is measured in volts (V).

A voltage of 1 V means 1 J of electrical energy is changing into other forms every time 1 C of charge passes through.

In this circuit:

- the charges have gained energy from the cell
- the lamp changes some of the electrical energy into light
- the resistor changes some of the electrical energy into heat
- the motor changes some of the electrical energy into kinetic energy.

Measuring voltage

A voltmeter is placed across (in parallel with) the component being measured, e.g. a battery.

Number of cells	Voltage (V)
1	1·5
2	3·0
3	4·5
4	6·0

A voltage reading of 4·5V here means 4·5J of energy is being supplied for each 1 C of charge passing through the battery.

Potential difference (or voltage) is a measure of the difference in energy (per unit charge) across two points in a circuit.

TOP TIP

Voltage is measured in volts with a voltmeter.

Conservation of energy

Energy is supplied as charges pass through a power supply. A voltmeter across the supply will measure the total energy supplied per unit of charge.

Energy is used as the charges move through the components or appliances in the circuit. Although each component may be using a different amount of energy, the total energy transferred out by the components is the same as the energy the source is supplying. This is in keeping with the Law of Conservation of Energy; energy is conserved.

Quick Test 24

1. What does charge experience in an electric field?
2. What do charges gain from a cell?
3. What units do we use to measure:
   (a) charge **(c)** energy
   (b) current **(d)** voltage?
4. What can the voltage across a component also be called?
5. The voltage on a battery is a measure of the energy given to the charges. True or false?
6. What is the symbol for a fuse?

Resistance and Ohm's Law

Physics to learn	Resistance, Ohm's law and the effect of temperature.
Revision guide	Describe resistance, investigate the relation between current, voltage and resistance, use graphs and calculate resistance.

Resistance

Materials can oppose the flow of charge through them. This is resistance.

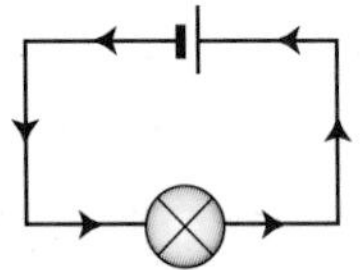

with no resistors in the circuit, there will be a large current

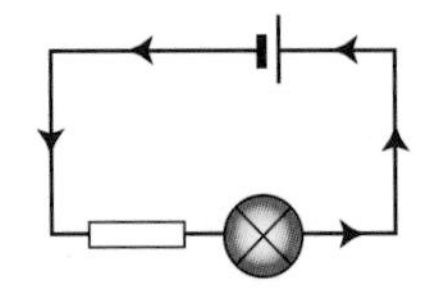

with a resistor in the circuit, there will be a smaller current

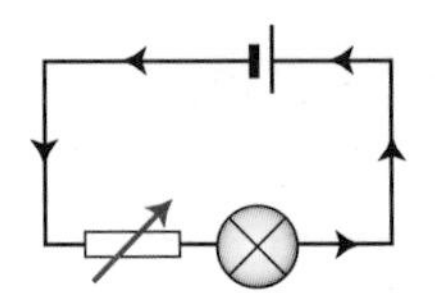

altering the value of this variable resistor changes the brightness of the lamp

Resistance is measured in ohms (Ω) using an ohmmeter. 1Ω is the resistance between two points on a conductor when a constant potential difference of 1V between them produces a current of 1A. Increasing the resistance of a circuit decreases the current in that circuit.

Resistance can be calculated using: $\textbf{Resistance} = \dfrac{\textbf{voltage}}{\textbf{current}}$ $R = \dfrac{V}{I}$

Example

A current of 3A is created in a circuit when a p.d. of 12V is applied across a motor.

Calculate the resistance of this motor. $R = \dfrac{V}{A} = \dfrac{12}{3} = 4\Omega$.

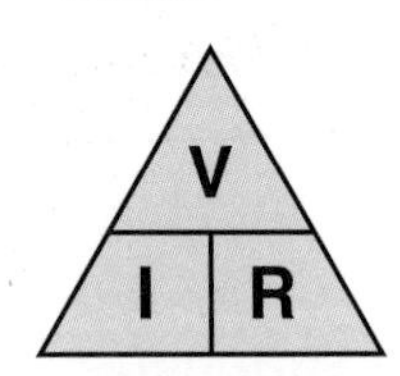

Ohm's law

To further investigate the relationship between voltage and current across a fixed resistance, we can apply a range of voltages or potential differences across a resistor and measure the dependant currents.

Plotting the results of the experiment, the graph reveals a straight line passing through the origin.

Note that to calculate resistance from the gradient that voltage must be on the y axis and current must be on the x axis. This may be the opposite from how you plotted the results after doing your experiment.

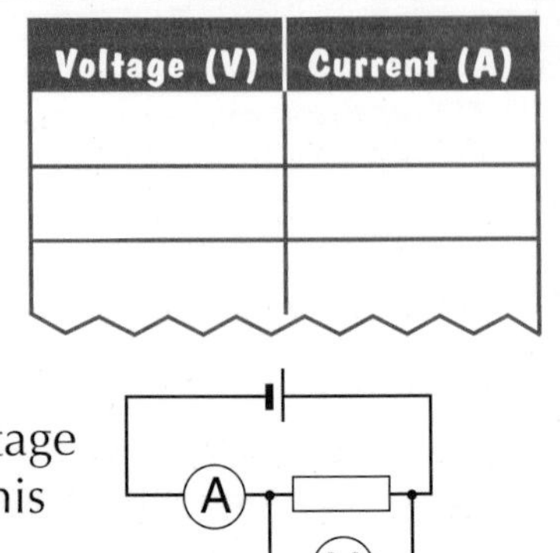

The shape of the graph shows the current in a resistor is directly proportional to the voltage applied.

This conclusion is known as Ohm's law.

The ratio $\frac{V}{I}$ remains constant for different currents.

$R = \frac{V}{I}$ $V = IR$

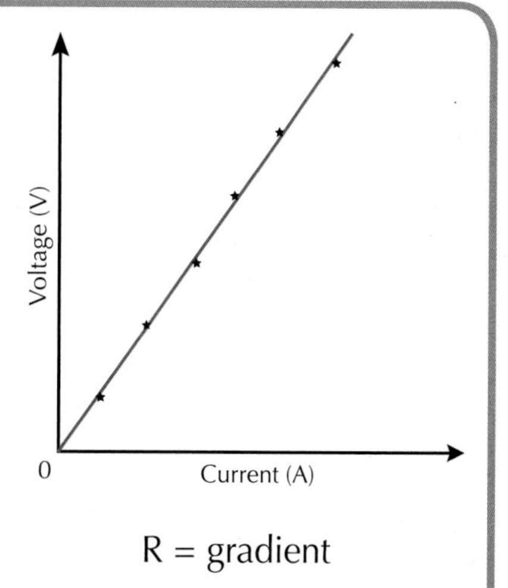

R = gradient

Example

What voltage is needed to have a current of 2 A in a circuit, when its resistance is 5 Ω?

$V = IR = 2 \times 5 = 10\text{V}$

Resistance and temperature

A pupil repeated the Ohm's law experiment on two different torch bulbs.

The two bulbs show different gradients as they have different resistances.

However, while the currents rise as the voltage increases, the currents do not continue to rise as much as expected.
Why not?

As the voltage increases, the current increases and the bulb filament increases in temperature. This, in turn, increases the resistance of the filament instead of remaining constant. The current no longer increases in proportion.

TOP TIP

Ohm's law is one of the most important equations in electric circuits, $V = IR$.

Quick Test 25

1. When resistance decreases, what does current do?
2. What meter can measure resistance?
3. Calculate the voltage required to create a current of 1 A in a 3 Ω resistor.
4. What assumption is made during the Ohm's law experiment?
5. A circuit resistance of 24 Ω is changed to 12 Ω. What will happen to the current?
6. A variable resistor has its resistance changed from 10Ω to 100Ω while supplied by a voltage of 25V. Calculate the minimum and maximum current.
7. Calculate the current drawn from the mains voltage of 230V by a component of resistance 1 kΩ. Calculate the current drawn if the component is used abroad with a voltage of 115V.

Components and symbols

Physics to learn	Practical circuit components and symbols.
Revision guide	Describe the function and application of electrical and electronic symbols.

Symbols

An electrical component is a part of a circuit. Circuit symbols are used when we draw electrical components in a circuit diagram.

Component	Symbol	Description
Cell		• A source of voltage and energy. • Used to provide a steady d.c. supply in low power devices, such as remote controls.
Battery		• A series of cells. • Increased power from a series of cells. • Batteries can be small or the size of a room. • Alessandro Volta invented the first battery.
Lamp		• Converts electrical energy to light energy. • Used as a light in an electrical circuit or an output device in electronic circuits.
Switch		• **Closed:** Makes a circuit, current exists, ideally it has no resistance when closed. • **Open:** Breaks a circuit, no current, ideally it has infinite resistance when open.
Resistor		• A component that opposes current in a circuit. • The resistor converts electrical energy to heat energy. • Many components also have resistance.
Variable resistor		• A resistor whose resistance can be varied to change current. • Variable resistors can be used as volume controls or dimmer switches.
Voltmeter	V	• Measures voltage or potential difference in volts. • Connect in parallel.
Ammeter	A	• Measures electrical current in amperes. • Connect in series.

Component	Symbol	Description
LED		• A light-emitting diode is often used as a low power output device to indicate on or off. • Recent developments mean we now have efficient LED lamps in the home.
Motor	M	• Converts electrical energy to kinetic energy. • Inside the motor electric current passes through a magnetic field to produce motion. • Motors are in everything from printers to washing machines.
Microphone		• Converts sound energy to electrical energy. • Often used as the input to an amplifier.
Loudspeaker		• Converts electrical energy to sound energy. • Often used as the output device after an amplifier.
Photovoltaic cell		• Also called a solar cell, it is an electrical device that converts the energy of light directly into electricity by the photovoltaic effect. • Solar cells can power satellites, offshore buoys and be a domestic source of energy.
Fuse		• A wire in the fuse melts when the current goes too high. • Protects the wiring from overheating. • It is important to match the fuse size to the power of the appliance in use.
Diode		• Only allows current in one direction. • Can protect a circuit from a power supply being connected the wrong way or change a.c. power to d.c.
Capacitor		• Stores charge and energy. • Used in audio electronics to provide energy for peaks in demand, for time delay circuits and for camera flashes, among many uses.
Thermistor		• A resistor whose resistance can be changed with temperature. • Makes a good temperature sensor as an input device in electronics.
LDR		• A resistor whose resistance can be changed with light. • Makes a good light sensor as an input device in electronics.

Quick Test 26

1. What is the symbol for an LED?
2. Find a device that could be an input sensor.
3. Find a device that could be an output sensor.
4. Draw symbols for components from memory.

Electronic circuits

Physics to learn	Transistor circuits.
Revision guide	Identify transistor symbols and their function in circuits.

Analogue and digital

Analogue signals are **continuously variable**. Most physical quantities, such as sound, heat and light, are analogue.

A microphone attached to an oscilloscope will display an analogue pattern with speech.

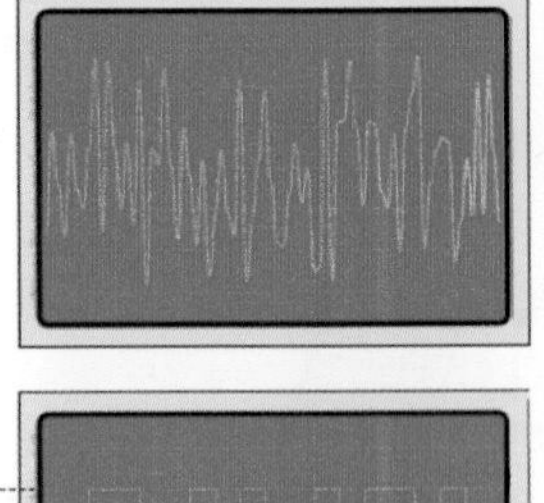

Digital signals have only two states. These are often called: **on/off, 5V/0V, high/low,** or **1/0**.

A CD player would output a digital pattern on an oscilloscope.

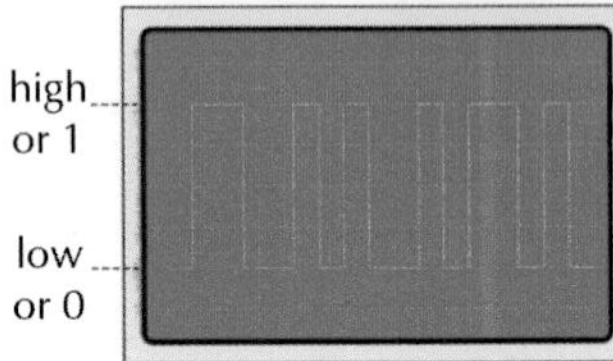

Input and output devices

Input	Output
Microphone Sound energy → electrical energy	**Loudspeaker** Electrical energy → sound energy
Solar cell Light energy → electrical energy	**Electric motor** Electrical energy → kinetic energy M
Thermistor As temperature increases, its resistance decreases to ohmmeter	**Relay** A relay is an electrically operated switch
Light dependent resistor (LDR) As light intensity increases, its resistance decreases to ohmmeter	**Light emitting diode (LED)** Electrical energy → light energy

Electronic circuits are made with INPUT → PROCESS → OUTPUT sub-systems.

The transistor will be our process sub-system.

The transistor

TOP TIP

Note, the transistor has three connections.

A transistor is an electronic switch. It is a voltage-controlled switch that responds to a change at the input.

This circuit shows an npn transistor.

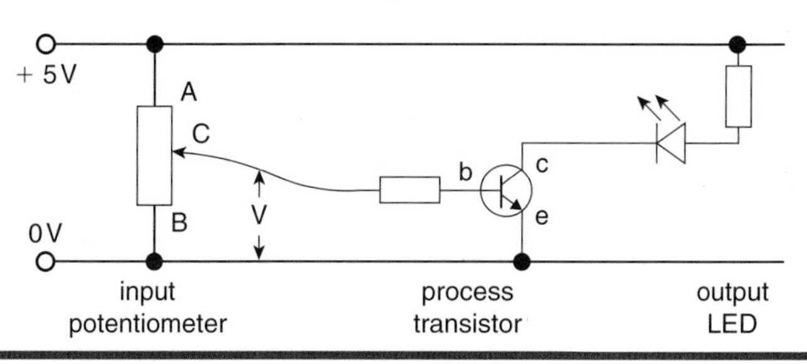

When input $V > 0{\cdot}7$V then the transistor conducts and the LED has been switched on.

When input $V < 0{\cdot}7$V or negative then the transistor does not conduct and the LED goes off.

Transistor circuits

TOP TIP

Remember the transistor symbol.

Light-controlled circuits

As the light increases, the resistance of the LDR decreases, the input voltage decreases below 0·7V and the transistor does not conduct so the LED goes off. If light decreases, the LED goes on.

A light increase could switch on a security warning light.

A light decrease could switch on a night light.

The variable resistor is adjusted to set the light level for the transistor switching on or off.

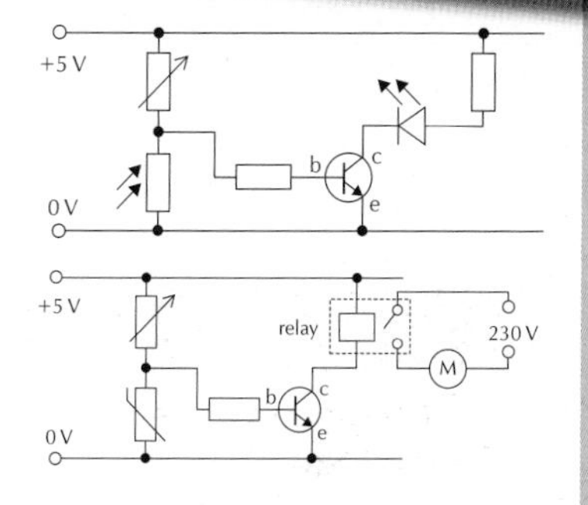

Temperature-controlled circuits

As temperature increases, the resistance of the LDR decreases, the input voltage decreases below 0·7V and the transistor does not conduct so the relay and motor go off. If temperature decreases, the relay and motor go on.

A temperature increase could switch on a fan.

A temperature decrease could switch on a heater.

TOP TIP

The MOSFET is another type of transistor

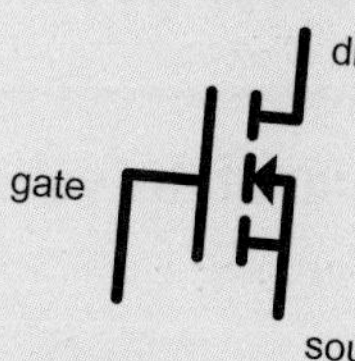

A MOSFET transistor will switch at a different voltage, say 2V.

Quick Test 27

1. Describe an analogue signal.
2. Describe a digital signal.
3. What is the main purpose of a transistor in electronics?
4. Why is a LDR in series with a resistor at the input stage?
5. What is the purpose of the variable resistor?

Series circuits

Physics to learn	Creating and measuring series circuits.
Revision guide	Calculating current, voltages and resistances in series.

Series circuits

TOP TIP

In a series circuit there is only one path for the flow of charge.

Series circuits have all the components in one loop.

There are no branches.

A series circuit is turned on or off by a single switch anywhere in the circuit. A break in the circuit at any point causes the whole circuit to stop working.

Calculating current

The current is the same at all points round a series circuit.

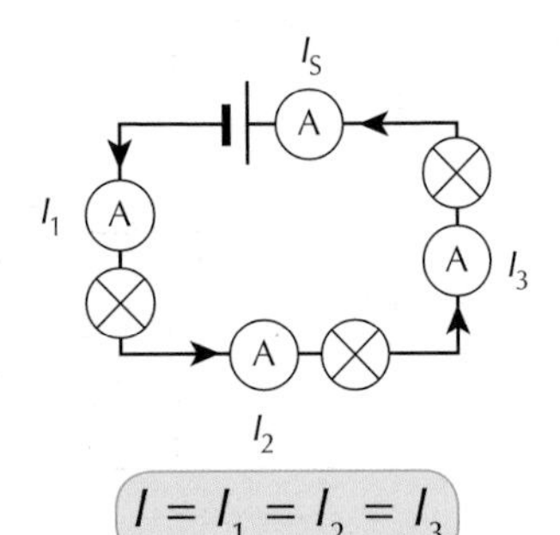

Charges flow round all points in the circuit at the same time.

Note how the ammeter is moved round the circuit to measure the current. It is always in line or in series in the circuit.

The ammeter readings will be the same all the way round.

Different circuits will have different readings but the values will still be the same all the way round.

Calculating voltage

The sum of the voltages across all the components in series is equal to the supply voltage.

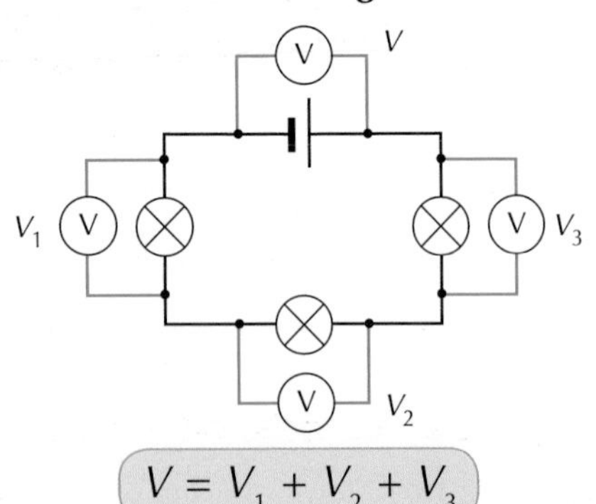

Note how the voltmeter is always placed across or in parallel with the component whose voltage is being measured.

The energy from the supply is shared out across the different components in the series circuit so each component gets a share of the voltage.

Each component will have a different voltage unless the components are identical.

Measuring resistance

TOP TIP

Resistance is measured with an ohmmeter, without the supply connected.

Resistance is the opposition to current and is measured in ohms (Ω).

The total resistance in series is equal to the sum of the individual resistances.

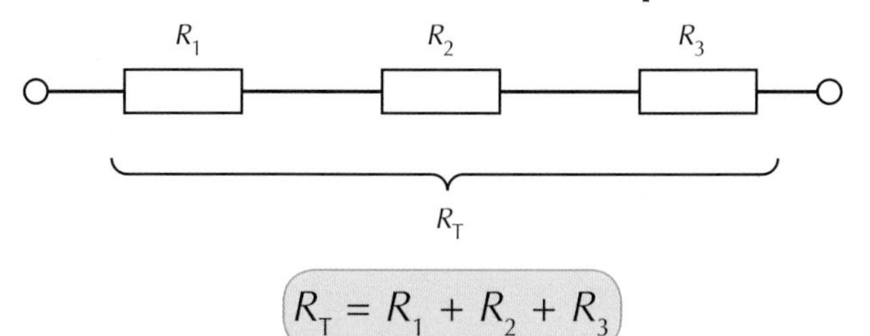

$$R_T = R_1 + R_2 + R_3$$

TOP TIP

In series, the total resistance will be larger than the largest resistor value.

If we join components in series, we increase the resistance of the circuit. The current will decrease.

Potential dividers

Potential divider circuits use two or more resistors in series or a potentiometer to provide a part of a supply voltage (V_s).

The voltage divides in the ratio of the resistors: $\frac{V_1}{V_2} = \frac{R_1}{R_2}$

$V_2 = \frac{R_2}{R_T} \times V_S$ or $V_2 = \frac{R_2}{R_1 + R_2} \times V_S$

$R_T = R_{Total}$ and $V_s = V_{supply}$

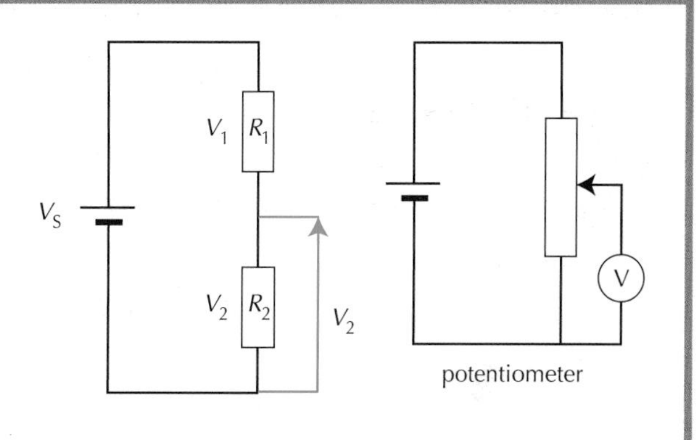

TOP TIP

When connecting components to an ohmmeter to measure resistance, the circuit power supply must be disconnected.

Quick Test 28

1. How are components connected in a series circuit?
2. 5, 10 and 15 ohm resistors are connected in series.
   (a) Calculate what voltage each gets from a 15V supply.
   (b) If the current at the 10 ohm resistor is 0·5 A, calculate the current at the other resistors.
   (c) Calculate the total resistance of these resistors in series.
3. A 6V battery is placed across three identical lamps in series. Calculate the voltage across the middle lamp.

Parallel circuits

Physics to learn	Creating and measuring parallel circuits.
Revision guide	Calculating current, voltages and resistances in parallel.

Parallel circuits

Parallel circuits have branches and junctions. There is more than one path for the charges to follow. A break in one branch has no effect on the other branches. There can be a switch for, and in, each branch and there can be a master switch beside the supply for the whole circuit.

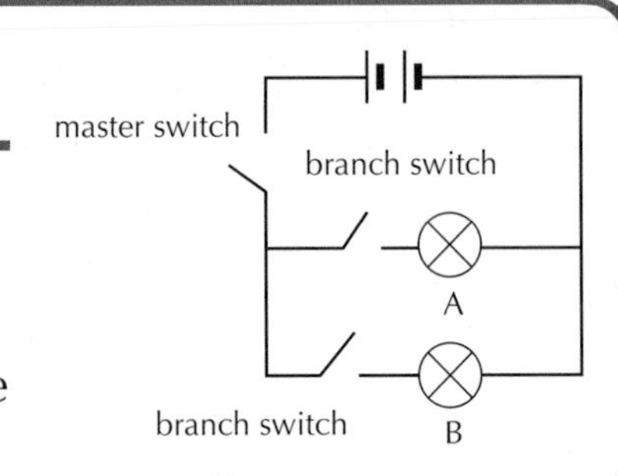

Calculating current

The circuit has different paths for charges to follow. **The current drawn from the supply depends on and is equal to the sum of the currents in the parallel branches.**

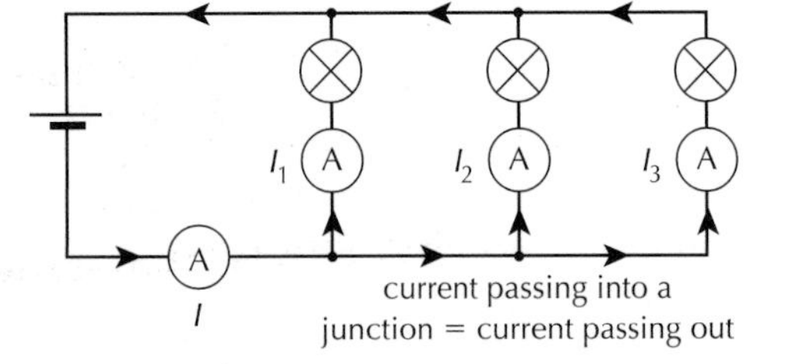

Note that I increases as more branches are added in parallel.

Each branch has its own current, independent of the currents in the other branches.

$$I = I_1 + I_2 + I_3$$

Calculating voltage

The voltages across components in parallel are the same and equal to their supply voltage.

$$V = V_1 = V_2 = V_3$$

The components can all be different yet they will all receive the same voltage as the supply voltage.

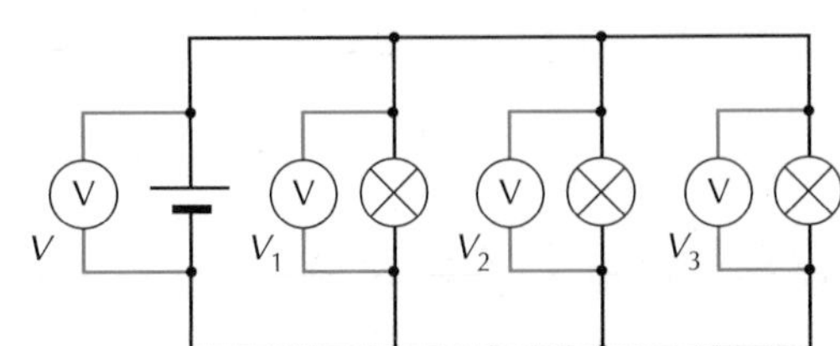

Conservation of energy

The energy supplied by the battery to each unit or coulomb of charge is equal to the energy given out by each unit or coulomb of charge. This is true for series or parallel circuits.

Calculating resistance

If we join components in parallel we decrease the resistance of the circuit. The current will increase.

The greater the number of branches, the smaller the total resistance and the greater the total current.

The total resistance of a parallel combination is calculated using a more complex formula.

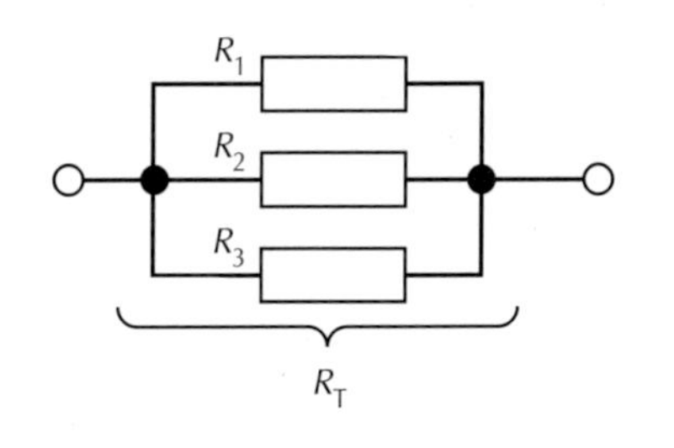

$$\frac{1}{R_T} = \frac{1}{R_1} + \frac{1}{R_2} + \frac{1}{R_3}$$

For example, if $R_1 = 2\,\Omega$, $R_2 = 3\,\Omega$, $R_3 = 4\,\Omega$

$$\frac{1}{R_T} = \frac{1}{2} + \frac{1}{3} + \frac{1}{4} \qquad \frac{1}{R_T} = \frac{6}{12} + \frac{4}{12} + \frac{3}{12} \qquad \frac{1}{R_T} = \frac{13}{12}$$

$$\frac{R_T}{1} = \frac{12}{13} \qquad R_T = 0\cdot92\Omega$$

TOP TIP

The total resistance of parallel branches is always smaller than the smallest branch resistance!

Resistor combinations

TOP TIP

The total of two identical resistances in parallel is equal to half of one of the resistances.

A circuit can have series and parallel combinations:

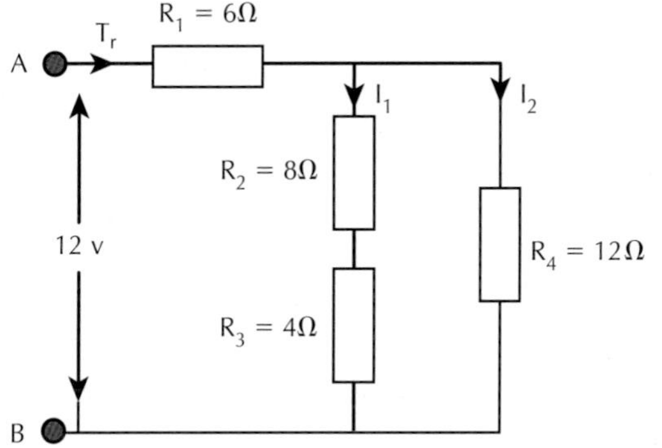

Both circuits are equivalent; they have a total resistance of 12 ohms.

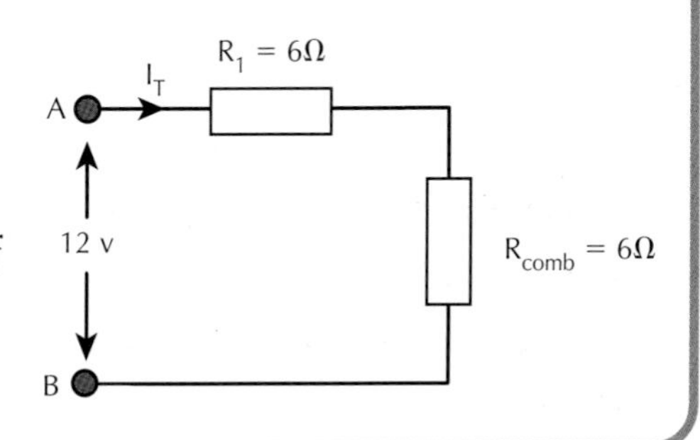

Quick Test 29

1. 5, 10 and 20 ohm resistors are connected in parallel. If the current at the 10 ohm resistor is 0·5A, calculate the current at the other resistors. Calculate the total resistance of these three resistors in parallel.
2. A 6V battery is placed across a motor, a heater and a bulb all in parallel. Calculate the voltage across the heater.

Energy and power

Physics to learn	Energy and power. Estimate power ratings and energy consumption.
Revision guide	Use the relationship between energy, power and time.

Energy

Around the house, circuits convert electrical energy into heat, light, kinetic, sound and potential energy.

Energy has the symbol E and is measured in joules (J).

Remember:

Kilojoule (kJ) = 1000 J (10^3 J)

Megajoule (MJ) = 1 000 000 J (10^6 J)

Gigajoule (GJ) = 1 000 000 000 J (10^9 J)

Terajoule (TJ) = 1 000 000 000 000 J (10^{12} J)

Power

Power (P) is measured in watts (W).

We can usually group the power of an appliance by the type of energy it produces. Appliances that transfer electrical energy to heat tend to be high power, whereas appliances that produce light from the electricity tend to be low power.

Appliance	Power rating
Cooker	12000W
Kettle	2000W
Iron	1900W
Computer	250W
TV	100W
Radio	20W
Light bulb	10W
Clock	10W

TOP TIP

Check the power of appliances in your home.

Energy, power and time

Power is the energy transferred or dissipated in unit time. The basic unit of time (t) in physics is the second (s).

A modern light bulb may have a power rating of 11 W. The light bulb converts 11 J of energy into light and a little heat every second.

Power = $\frac{\text{Energy}}{\text{time}}$ $P = \frac{E}{t}$

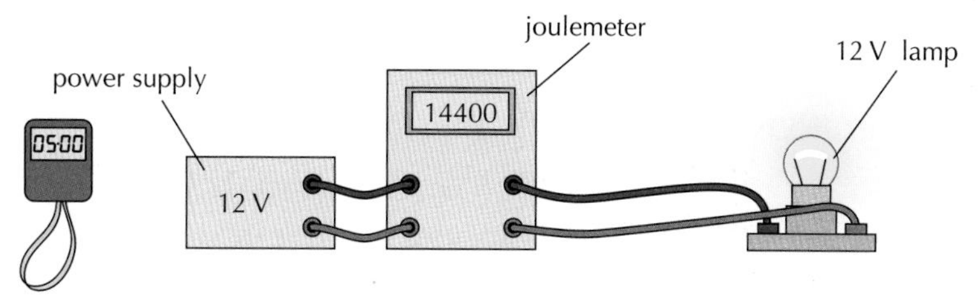

Energy can be measured directly with a joulemeter.

Time can be measured with a stop-clock.

So, for this lamp: $P = \frac{E}{t} = \frac{14400}{300} = 48\,W$

How many joules of energy have been converted?

To calculate the total amount of energy an appliance has converted we use the equation: Energy = Power × time or $E = Pt$

Energy consumption and energy meters

The joule is a very small unit of energy and electrical bills measure energy used in 'units'. To calculate units, power is left in kilowatts and time is kept in hours. 1 kW for 1 hour gives 1 unit of energy. A unit is 1 kilowatt hour.

Example

Calculate the energy used when a 2 kW fire is turned on for 6 hours. The cost of each unit is 15p.

$E = Pt = 2 \times 6 = 12$ kWh or 12 units.

Cost: $12 \times 15 = 180$p = £1·80

Quick Test 30

1. Name **three** high power appliances.
2. Calculate how much energy a 3 kW fire uses in 6 hours in joules.
3. An LED emits 1·2 J of light in 1 minute. What is its power rating?
4. How many joules of energy are there in 1 kilowatt-hour?
5. At a cost of 15p per unit, calculate the cost of:
   (a) a 14 W bulb on for 10 hours.
   (b) a 3000 W fire turned on for 5 hours.
   (c) a 2000 W tumble dryer used for 1 h 30 mins.

Power and fuses

Physics to learn	Power equations and fuse values.
Revision guide	Use appropriate relationships and fuses.

Power equations

Power (P) is the energy (E) transferred in unit time (t) and is measured in watts (W).

As power rating increases, current increases. Power also depends on voltage. If the voltage supplied to a component in a circuit, e.g. a lamp, is decreased, the lamp will dim as it is now working at a lower power. In fact, the power rating can be confirmed to be equal to the product of voltage and current.

Power = current × voltage $P = IV$

If a voltage of 1 V across a component creates a current of 1 A, then the power is 1 W, and 1 J of energy will be transferred in a time of 1 s.

More electrical power equations

If we substitute Ohm's law into our new power equation we can derive more:

Substituting $V = IR$ into $P = IV$, $P = IV = I(IR) = I^2R$

Substituting $I = \frac{V}{R}$ into $P = IV$, $P = IV = \left(\frac{V}{R}\right)V = \frac{V^2}{R}$

TOP TIP

We now have four equations we can use with electrical power:

$$P = \frac{E}{t}$$

$$P = IV$$

$$P = I^2R$$

$$P = \frac{V^2}{R}$$

Transmission of power

Electrical energy is carried in long transmission lines between the power station and our homes or industries. These lines make up part of the National Grid.

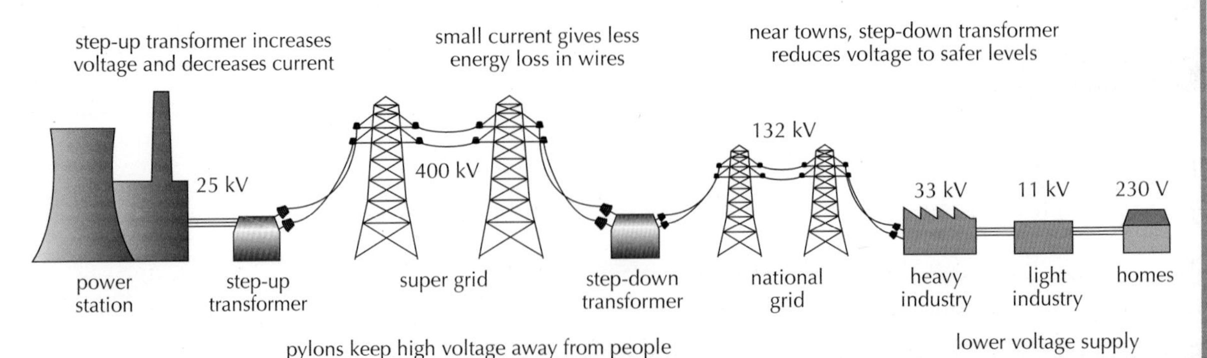

Power loss

Long transmission lines are made of low resistance cable, however, there is power loss in the lines and electrical energy will change to heat. For example, power is sent down transmission lines at the generated voltage of 25 kV, and 100 A of current is drawn through lines of resistance of 6 Ω.

The power sent, P = IV = 100 × 25 000 = 2 500 000 W or 2·5 MW.

$P_{loss} = I^2R$

The power loss depends on the square of the current.

$P = I^2R = 100^2 \times 6 = 60\,000$ W or 60 kW

Plugs and fuses

Fuses are fitted to plugs to protect the flex, which can overheat if too large a current is drawn. The fuse may protect the appliance.

We know that mains voltage is 230 V. We check the power rating of an appliance then calculate what current will exist, e.g. for a heater rated at 2 kW:

$$I = \frac{P}{V} = \frac{2000}{230} = 8{\cdot}7\text{A}$$

A fuse higher than the calculated current should be chosen. A 13A fuse should be used.

Quick rule

As it is very important that the correct value of fuse is selected for a plug, a quick rule has been derived for safety:

A 3A fuse should be selected for most appliances rated up to 720W

A 13A fuse should be selected for appliances rated over 720W

This low power desk lamp is fitted with a 3A fuse

This high power heater is fitted with a 13A fuse

Quick Test 31

1. A drill uses 90 000 J of energy in 3 minutes. Calculate its power.
2. A 1·4 kW vacuum cleaner is used for 30 minutes. Calculate the energy it used.
3. A 3V bulb draws 250 mA from a battery. What is its power? Calculate the energy used in 5 minutes.
4. A heater draws a current of 6 A through its 40 Ω element. Calculate its power rating.
5. Calculate the resistance of the filament of a 60 W, 230V lamp.
6. Show that P = IV, $P = I^2R$ and $P = \frac{V^2}{R}$ are equivalent.

Heat

Physics to learn	Heat, temperature, kinetic theory and storing energy.
Revision guide	Distinguish between heat and temperature. Use of materials in storing energy.

Kinetic theory

Atoms or molecules are always in motion. Some are free to move and some are simply vibrating. The movement of these particles depends on the amount of energy they have. Explaining or describing quantities such as pressure, temperature and temperature in terms of the motion of these particles is known as kinetic theory.

When we heat a substance, energy is being transferred to the atoms or molecules. These particles will increase their speeds. We say they have an increase in kinetic energy.

When a heater supplies energy to a liquid, the kinetic energy of the particles will increase and these particles will move about faster. As temperature varies with the average kinetic energy of these particles we see an increase in temperature.

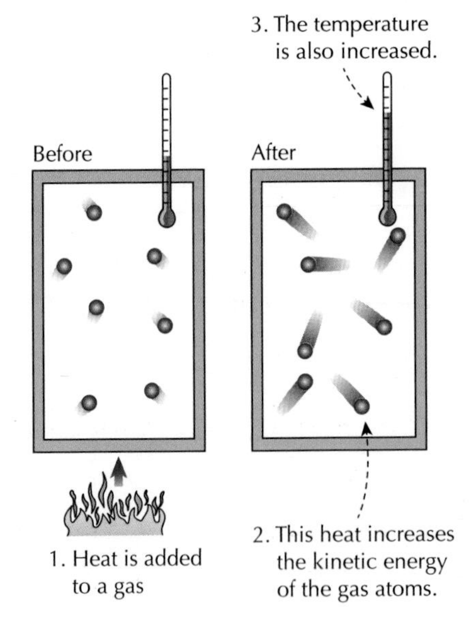

The temperature of a substance is a measure of the mean kinetic energy of its particles.

TOP TIP

Heat has the unit J
Temperature has the unit °C

Storing energy

When heat energy is added to a material, it is stored in the material.

This effect is used in storage heaters, which store heat in an insulated container of hot water or bricks, using cheap electricity during off-peak times. These materials store a lot of heat and this stored heat can then be gradually released into a home during the day.

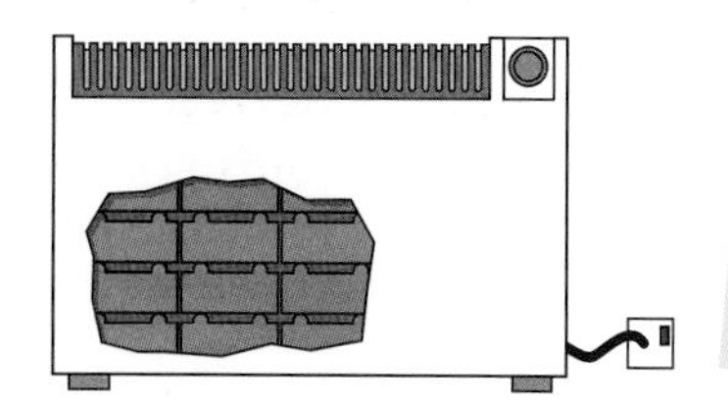

TOP TIP

Investigate new heating systems such as heat pumps and ground storage systems.

Heat and temperature

TOP TIP

Another unit used in physics for measuring temperature is the Kelvin scale. Where Celcius takes the freezing point of water as 0° the Kelvin scale takes the lowest possible temperature – absolute zero – as 0 (-273°C).

Temperature

Temperature (T) is a measure of how hot a substance is and it is measured in degrees Celsius (°C).

Some typical temperatures	
Absolute zero	–273°C
North Pole	–60°C
Melting ice	0°C
Room temperature	20°C
Core body temperature	37°C
Hot tap water	50°C
Boiling water	100°C
Sun	10^6 °C

TOP TIP

Temperature => hotness, heat => energy.

Heat

Heat, E_h, is a form of energy and is measured in joules (J).

Heat energy can transfer by conduction, convection or radiation.

Heat will flow from hot objects to colder surroundings. The greater the temperature difference with the surroundings, the greater the rate of heat loss.

Quick Test 32

1. What quantity does temperature measure?
2. What quantity does heat measure?
3. Name **three** methods of heat transfer.
4. What is studied in kinetic theory?
5. What is stored when a material is heated?

Heat capacity

Physics to learn	Heat capacity and specific heat capacity.
Revision guide	Can investigate relationships with heat, can measure and do calculations with specific heat capacity.

Heat capacity

Heat capacity is the amount of heat energy required to raise the temperature of a substance. It depends on three quantities: temperature change, the mass of the substance and the substance's material.

The greater the temperature change desired, the more energy will be required.

A greater mass of material will require a greater amount of heat for the same temperature rise.

Different materials require different quantities of energy to raise their temperature by one degree.

Temperature change

The quantity of heat energy required to raise the temperature of a substance varies with the change in temperature required ($E \propto \Delta T$).

Mass change

The quantity of heat energy required to raise the temperature of a substance varies with the mass of the subsitance. ($E \propto m$).

Material change

The quantity of heat energy required to raise the temperature of a substance varies with the specific heat capacity of the material ($E \propto c$).

Energy (J) $\propto$ specific heat capacity ($J\,kg^{-1}\,°C^{-1}$).

Combining experiments shows heat energy depends on all three quantities:

$$E = cm\Delta T \qquad c = \frac{E}{m\Delta T} \qquad m = \frac{E}{c\Delta T} \qquad \Delta T = \frac{E}{cm}$$

TOP TIP

α = 'varies with' = 'is directly proportional to'

Example

The energy required to raise the temperature of 3 kg of water from 20°C to 50°C is:

$E = cm\Delta T = 4180 \times 3 \times 30 = 376\ 200\,J$.

Specific heat capacity

Specific heat capacity (c) is the quantity of energy required to change the temperature of 1 kg of mass of a substance by one degree Celsius.

This is also the energy 1 kg of a substance can store for each degree.

Water has a very high specific heat capacity:
$c = 4180\,J\,kg^{-1}\,°C^{-1}$.

A lot of heat is needed to make water hot. Water can store a lot of heat.

TOP TIP

The same quantity of energy used to heat up a substance will be given out when the substance cools.

Measuring specific heat capacity

This block of steel has holes drilled to take the heater and the thermometer. Oil conducts the heat. To measure the specific heat capacity:

- measure the **energy** supplied
- measure the mass of the block and the **temperature** rise
- then **calculate** using $c = \frac{E}{m\Delta T}$

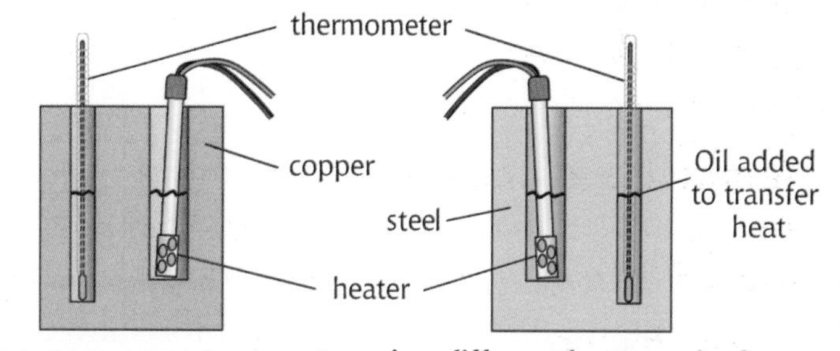

Same mass, same temperature rise, different heat required:
Copper, $c = 386\ J\ kg^{-1}\ °C^{-1}$ **Steel, $c = 902\ J\ kg^{-1}\ °C^{-1}$**

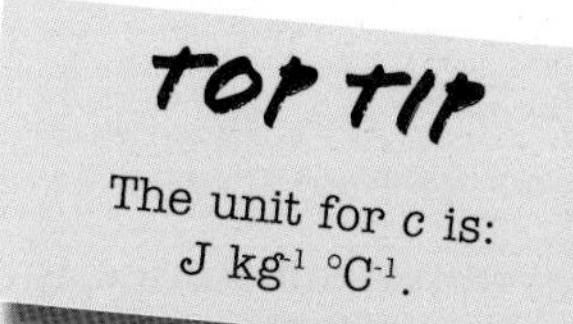

Quick Test 33

1. If you heat 1 kg of water in a kettle from room temperature of 20°C to boiling point of 100°C, how much energy does this take?
2. Calculate how long this kettle would take to boil if its power is 2200W.
3. In practice, this kettle took 3 minutes. Why?
4. What is the unit of specific heat capacity?
5. A kettle gives out 167 200J of heat energy when water at 100°C cools to 20°C. Calculate how much water is in the kettle.

Latent heat

Physics to learn	Latent heat and specific latent heat.
Revision guide	Latent heat of fusion and vaporisation.

Change of state

The three main states of matter are solid, liquid and gas.

Solid state: Atoms are fixed in a structure but have energy to vibrate.
Liquid state: Atoms have more energy and are free to tumble.
Gas state: Atoms have enough energy to break free and stay further apart.

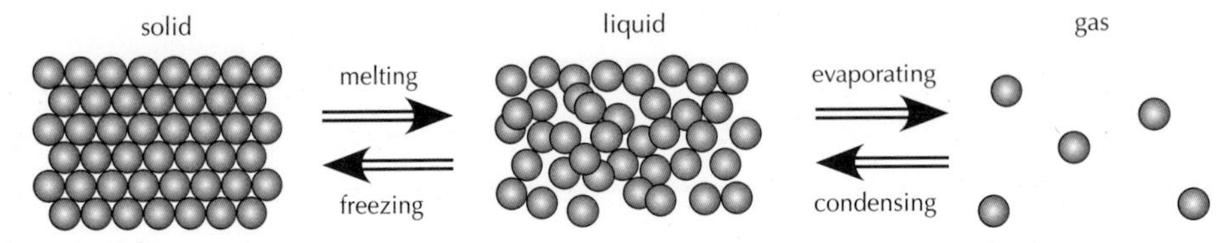

When a solid changes to a liquid this is called melting and energy is put in.
When a liquid changes to a gas this is called evaporation and energy is put in.
When a gas changes to a liquid this is called condensation and energy is given out.
When a liquid changes to a solid this is called freezing and energy is given out.

TOP TIP

When ice is melting, the ice is 0°C and the water is 0°C.

Energy has to be gained or lost by a substance to change its state. Time is taken for a substance to change state. During this time there is NO change in temperature.

A picnic box cool pack is frozen in a fridge. When placed beside the food it takes heat energy from the food to melt, keeping the food cool. When we step out of the shower, water on our skin evaporates, taking heat energy from our bodies, making us cool. In these examples, a change of state causes a change in temperature!

Cooling curves

When a substance is at higher temperature than its surroundings, energy is given out.

When the temperature drops, energy is being given out. Note: The temperature drops quickest at the start when there is a large temperature difference with the surroundings. The temperature finally stops dropping when it is at the same temperature as the surroundings (room temperature).

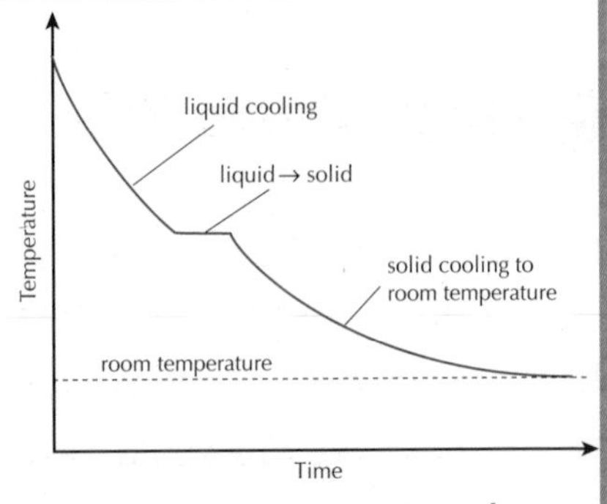

The temperature stays the same during a change of state. The temperature stays the same, but energy is still being given out! The temperature stays the same until all the atoms have changed state. Each atom loses some energy.

The amount of heat energy required to change the state of a substance:

- varies with the mass being heated ($E \propto m$);
- varies with the specific latent heat l of the material being heated ($E \propto l$).

Note: A different mass of material will require a different amount of heat to change its state. The same mass of different materials requires different quantities of energy to change their states.

The amount of heat energy depends on these two factors:

$E = ml$ (E in J, m in kg, l in J/kg) $l = \frac{E}{m}$ $m = \frac{E}{l}$

TOP TIP

Energy can cause a change of state with NO change in temperature.

Specific latent heat

The specific latent heat, l, is the amount of heat energy to change the state of 1kg of a substance.

Latent heat of fusion – the amount of energy taken in or released to change between a solid and a liquid state.

e.g. specific latent heat of fusion of ice, $l = 3{\cdot}34 \times 10^5$ J/kg.

Latent heat of vapourisation – the amount of energy taken in or released to change between a liquid state and a gas.

e.g. specific latent heat of vaporisation of water, $l = 2{\cdot}26 \times 10^6$ J/kg.

TOP TIP

The specific latent heat is the amount of energy to change the state of 1 kg of a material.

Energy changes

The principle of conservation of energy allows us to calculate the heat energy supplied to a substance from an electrical heater.

Heat energy gained = Electrical energy used. $cm\Delta T = ItV$ or $ml = ItV$

Quick Test 34

1. If you heat 1 kg of water in a kettle from room temperature of 20°C to boiling point of 100°C, how much energy does this take?
2. How long would this kettle take to boil if its power is 2200W?
3. In practice, this kettle took 3 minutes. Why?
4. After boiling the water, this kettle is left on. How much more energy is used while all the water turns to steam?
5. How long does the water take to evaporate?

Pressure

Physics to learn	Pressure, force and area.
Revision guide	Explain pressure on a surface, calculate and measure pressure.

Pressure

Pressure is a measure of force on unit area. The combination of a large force on a small area exerts high pressure and is most likely to penetrate a surface. A small force on a large area exerts low pressure.

Examples of high pressure

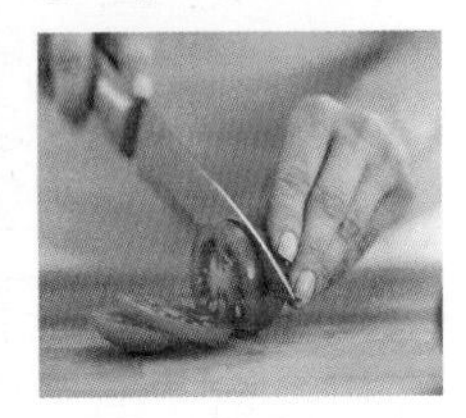

Examples of low pressure

Pressure and kinetic theory of gases

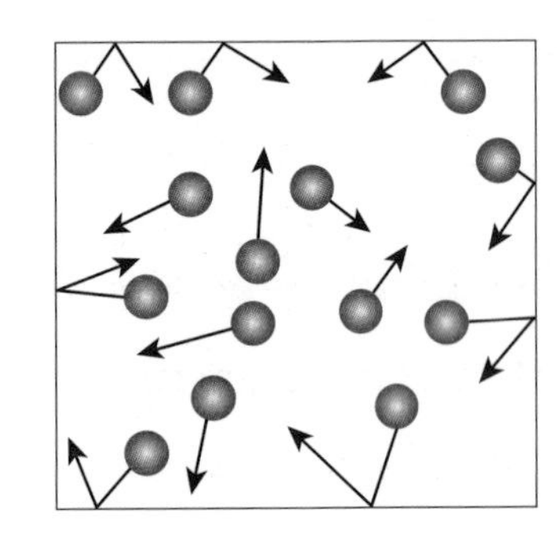

Air particles exert pressure. Air pressure is caused by the weight of air above exerting force on the air molecules at ground or sea level.

When we examine gas particles in a container we see that they are in constant motion. They have kinetic energy. They collide with each other but also against the walls of the container. Because of the motion, each molecule exerts a tiny force but all the particles exert a larger average force.

The force per unit area is the gas pressure.

Force is measured in newtons (N) and area in square metres (m^{-2}) so pressure can be measured in Nm^{-2}.

As more gas is pumped into the container, the number of moving particles increases, the number of particle collisions increases, the average force increases on the same area, and the pressure increases.

TOP TIP

Lay a box on its side and there will be less pressure than if it is stood on end.

Calculating pressure

Pressure is the force per unit area, when the force is pushing on a surface.

$$\textbf{pressure} = \frac{\textbf{force}}{\textbf{area}} \qquad p = \frac{F}{A}$$

Units: Nm^{-2} or pascal (Pa). The pascal is a unit of pressure.

When 1 newton exerts a force on 1 square metre, the pressure is 1 pascal.

$1\ Pa = 1\ Nm^{-2}$

Example

A person whose mass is 80 kg stands on one foot. The foot has an area of 200 cm^{-2}. Calculate the pressure exerted on the floor.

$$200\,cm^{-2} = 2 \times 10^{-2}\,m^{-2} \qquad p = \frac{F}{A} = \frac{mg}{A} = \frac{80 \times 9{\cdot}8}{2 \times 10^{-2}} = 3{\cdot}9 \times 10^{4}\,Pa$$

Measuring pressure

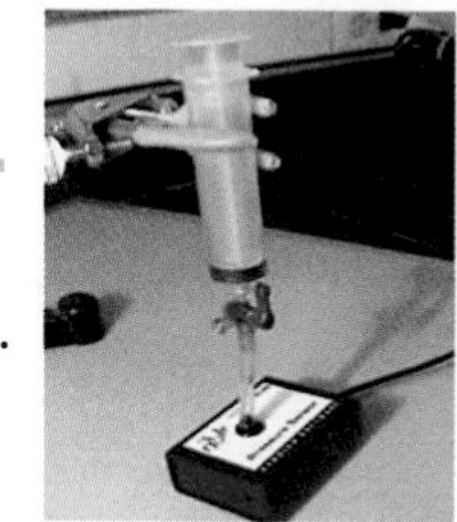

Weights are added to the piston at the top of the syringe. The weights exert force on the area of piston in contact with the gas in the syringe.

The additional pressure exerted on the gas can be calculated from:

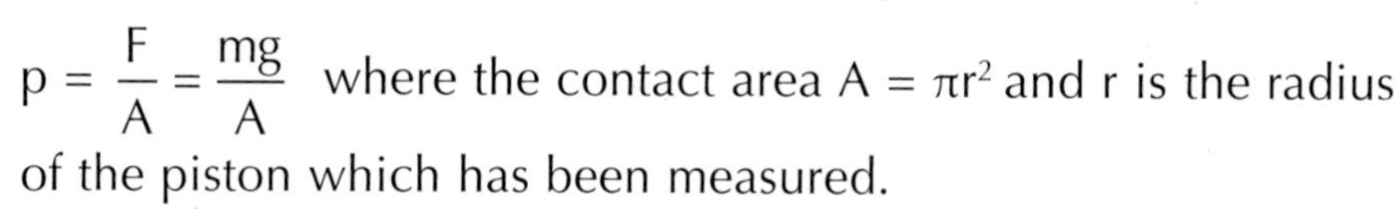

$p = \frac{F}{A} = \frac{mg}{A}$ where the contact area $A = \pi r^2$ and r is the radius of the piston which has been measured.

As the weights are added the change in pressure recorded by the sensor is seen to increase and the pressure values are shown to be equal to the calculated values of $\frac{F}{A}$, confirming the relationship.

TOP TIP

Atmospheric pressure is 101 000 Pa at sea or ground level.

Quick Test 35

1. A box is stood on end so that its contact area has been halved. What has happened to the weight and pressure on the floor?
2. Using kinetic theory, explain how pressure arises in a gas.
3. A box has a weight of 250 N and a contact area of 0·05 m^{-2}. Calculate the pressure on the floor.
4. An elephant has a mass of 7500 kg and stands on all four feet. If each foot has an area of 1000 cm^{-2}, what pressure is exerted?
5. If the same elephant stood on one foot, what would be its pressure?

Gas: Pressure - temperature

Physics to learn	The pressure law with its kinetic theory.
Revision guide	Describe and explain the effect of temperature on pressure. Derive and use the Kelvin scale.

Pressure law

Fixed volume and mass of air is trapped in a flask, and the flask is placed in water.

The water is heated and stirred.

The temperature in °C is measured with a thermometer or temperature sensor. The pressure in Pa or kPa is measured with a bourdon gauge or pressure sensor.

As the temperature rises, the pressure rises. There is a linear relationship between the temperature and the pressure. However, at 0°C there is still pressure as the line of best fit does not go through the origin.

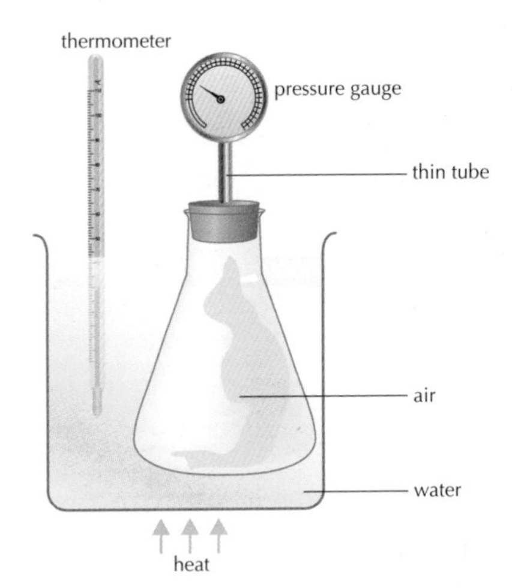

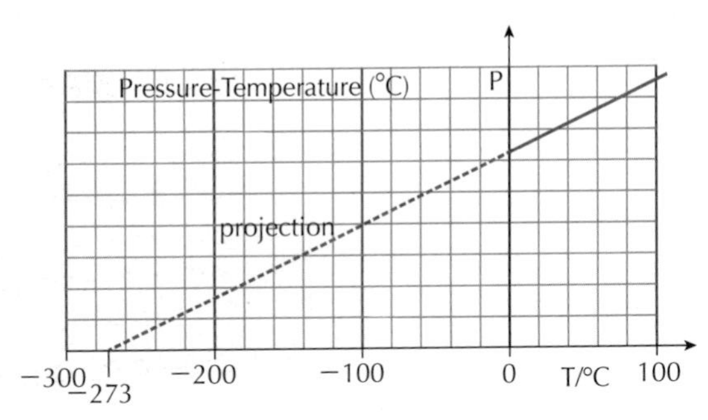

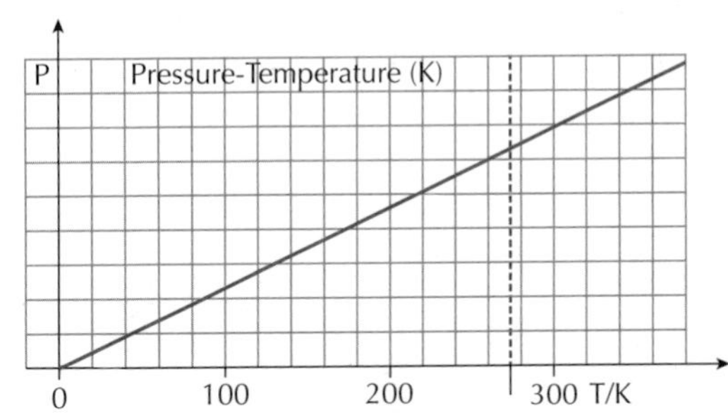

Only if we extend the line back to where there is zero pressure will there be a true zero of temperature. From the experiment, zero pressure is predicted at −273°C. Zero pressure indicates the true zero of temperature, called absolute zero. We need to change from the Celsius scale to the Kelvin scale.

Absolute zero is 0 K. The Kelvin scale uses the same division size as the Celsius scale.

On the Kelvin scale, pressure is directly proportional to temperature: $p \propto T$.

$$\frac{p}{T} = k$$ where k is a constant. $$\frac{p_1}{T_1} = \frac{p_2}{T_2}$$

This is known as the pressure law.

TOP TIP

Only with temperature in Kelvin will this gas law work!

TOP TIP

0 K = -273°C

GOT IT? ☐ ☐ ☐

Example

A sealed flask of gas has a pressure of 5×10^5 Pa at a temperature of 27°C. If the temperature drops to –27°C, what is the new pressure?

$p_1 = 5 \times 10^5$ Pa

$T_1 = 27°C = 300$ K

$T_2 = -27°C = 246$ K

$$\frac{P_1}{T_1} = \frac{P_2}{T_2} \qquad \frac{5 \times 10^5}{300} = \frac{P_2}{246} \qquad p_2 = 4{\cdot}1 \times 10^5\ Pa$$

Kinetic theory: the pressure law

If the temperature increases, the particles have more kinetic energy and the velocity increases. The particles hit the container walls more often and with greater force. Area (A) remains constant but average force (F) increases so the pressure (p) increases: $p = \frac{F}{A}$.

Temperatures

What is double the temperature of 20°C? We cannot just double 20!
First change to the Kelvin scale: 20°C = 293 K.
Now double, which makes 586K.
To return to Celsius, subtract 273.
This gives 313°C.

Temperature scales

Only the Kelvin scale starts from an absolute zero of temperature.

- Degrees Celsius to Kelvin: add 273.
- Kelvin to degrees Celsius: subtract 273.

TOP TIP

A temperature change of 100°C = a temperature change of 100 K. The divisions are the same size.

Quick Test 36

1. Air at 5°C and $1{\cdot}25 \times 10^5$ Pa is heated by the Sun to 25°C. Calculate the new pressure.
2. What assumptions did you make in Q1?
3. What is double the temperature of –10°C?
4. What four things about the particles increase with temperature?

Gas: volume - temperature

Physics to learn	Charles' law with its kinetic theory.
Revision guide	Describe and explain the effect of temperature on volume.

Charles' law

Air is trapped in a thin glass rod by a bead of mercury. Its mass is now fixed. As the bead is free to move, and one side is open to the atmosphere, the pressure of the trapped gas will always be the same as the pressure of the atmosphere. We can say that during the experiment there is a gas with a fixed mass and a constant pressure.

Bead
Ruler
trapped air
Water bath
Heat

This experiment considers the relationship between temperature and volume.

The water is slowly heated.

The water is also stirred so that the trapped gas is heated evenly and we can assume that it is at the same temperature as the water.

The temperature in °C is measured with a laboratory thermometer or temperature sensor.

The volume is found from a scale on the rule behind the rod. (You may be aware that if the glass rod has a constant width that the volume of the trapped gas is proportional to its length, i.e. if the volume doubles then the length will also show double, so we do not need to actually calculate the volume, we can just use the length.)

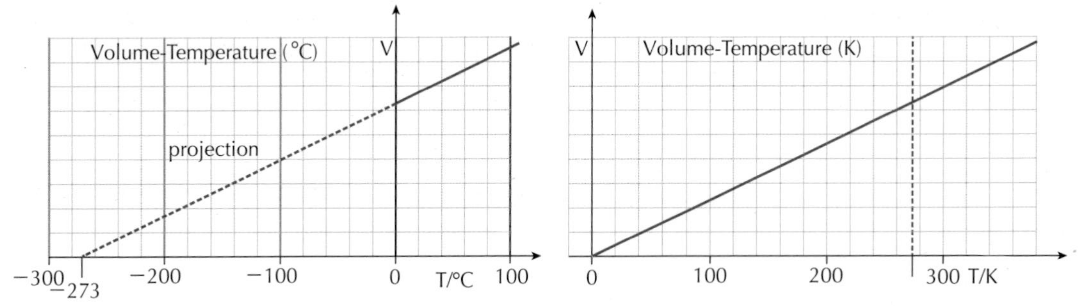

As the temperature increases, the volume increases. We can see there is a linear relationship between the temperature and the volume. However at 0°C there is still volume as the line of best fit does not go through the origin. We do not yet have a directly proportional relation between volume and temperature.

Only if we extend the line back to where the graph indicates zero volume will we see a true zero of temperature. From the graph, zero temperature is predicted at –273°C. Zero volume indicates the true zero of temperature, called absolute zero. We need to change from the Celsius scale to the Kelvin scale.

From experiments, we see that absolute zero is indicated where there is no pressure and no volume. The motion of all particles will have stopped.

Absolute zero is 0K. We can use the same divisions as the Celsius scale but we need to start with 0K at –273°C.

Now on the Kelvin scale, a true scale for temperature, volume is directly proportional to temperature: V α T.

$V \alpha T$ $\quad \frac{V}{T} = k \quad \frac{V_1}{T_1} = \frac{V_2}{T_2}$

This is **Charles' law**.

TOP TIP

Remember temperatures in Kelvins.

Kinetic theory

TOP TIP

When discussing kinetic theory you should include the word particles or molecules in your answer.

When we discuss pressure, volume and temperature we are looking at the large scale or "macroscopic" characteristics of the gas. These can be explained by discussing the behaviour of the particles at the small scale or "microscopic" level. These discussions, where we explain what is happening to the particles are referred to as Kinetic Theory.

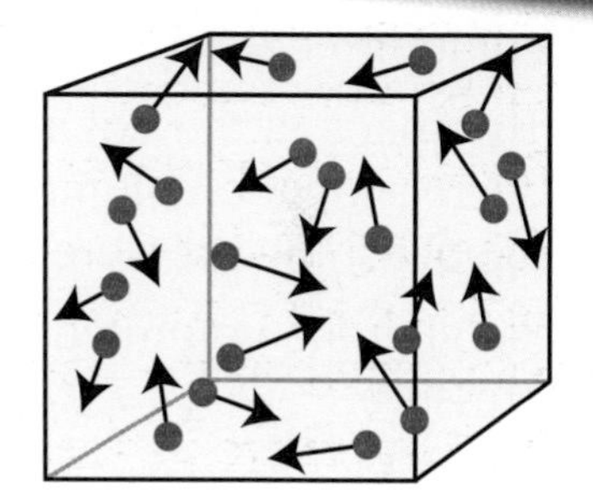

According to the Kinetic Theory, gases are made up of tiny particles in random, straight line motion. The particles move rapidly and continuously and make collisions with each other and the walls.

Kinetic theory: Charles' law

If the temperature increases, the particles have more kinetic energy and the velocity increases. The particles hit the container walls with more force. The volume increases so the surface area increases to keep the pressure constant.

The pressure, p, is constant: $p = \frac{F}{A}$. F increases and A increases.

Example. At what temperature will a litre of trapped gas be doubled in volume if it started at 0°C?

0°C = 273 K. $\frac{V_1}{T_1} = \frac{V_2}{T_2} \quad \frac{1}{273} = \frac{2}{T_2} \quad T_2 = 546K$

New temperature = 546 K or 273°C.

Quick Test 37

1. At 25°C, the volume of a fixed mass of gas is 3 litres. Calculate its volume at 125°C.
2. If a bubble of gas has a volume of 15 cm^3 at a room temperature of 5°C, why will it not double to 30 cm^3 at a temperature of 10°C?
3. What remains constant in the Charles' law experiment?

Final gas laws

Physics to learn	Boyle's law with its kinetic theory, the general gas equation.
Revision guide	Explain and calculate the effect of changing volume on pressure, with or without temperature change.

Boyle's law

A quantity of air, with a fixed mass and constant temperature, is trapped in a column.

Pressure is measured with a bourdon gauge or a pressure sensor connected to a computer. The volume is read from a scale on the column.

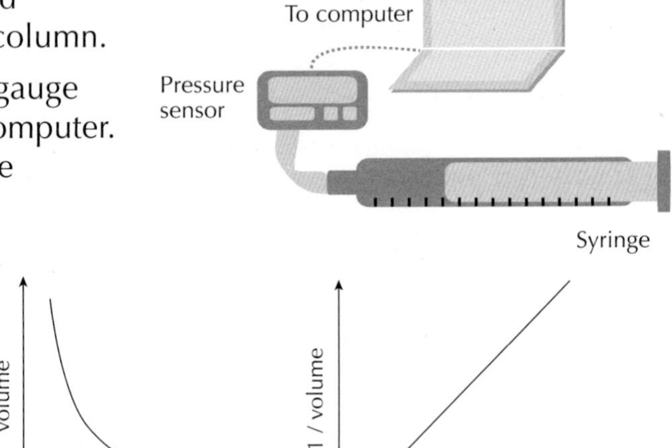

The space in the column is decreased, decreasing the volume. As this happens, the pressure increases.

Halving the volume doubles the pressure.

A graph of pressure against volume is not a straight line but indicates an inverse relationship.

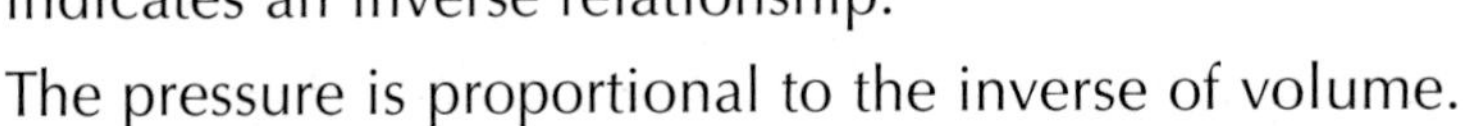

The pressure is proportional to the inverse of volume.

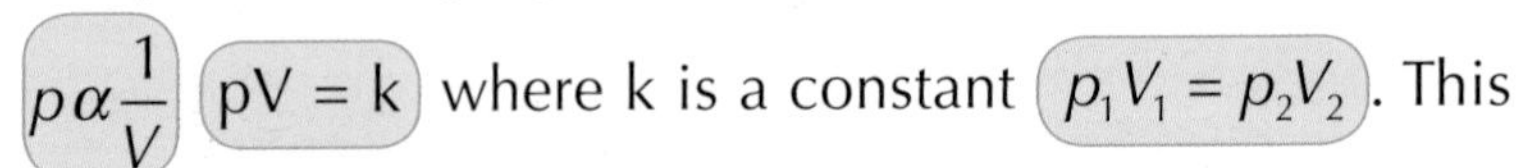

$p \alpha \frac{1}{V}$ pV = k where k is a constant $p_1V_1 = p_2V_2$. This is Boyle's law.

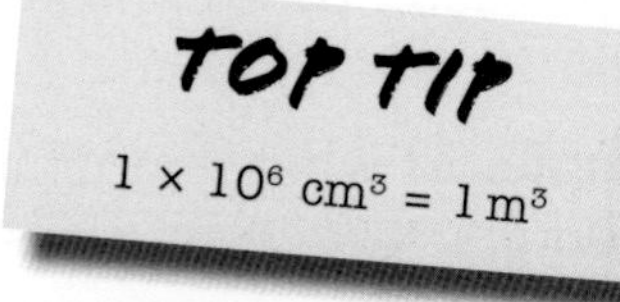

TOP TIP

1×10^6 cm³ = 1 m³

TOP TIP

Various units can be used in Boyle's law if they are the same units on both sides.

Kinetic theory: Boyle's law

As the temperature is constant, the kinetic energy and the velocity are constant.

If the volume is increased, the particles hit the container walls less often, therefore they exert less average force on the increased area of the container walls.

The pressure (p) decreases as force F decreases and area A increases:

$$p = \frac{F}{A}$$

General gas equation

There are times when all three of the quantities – pressure, volume and temperature – all change together. The three gas laws that we have found from experiments can be combined into one general gas equation. P, V, and T can all change.

Experiment 1: **Pressure law** $\frac{p_1}{T_1} = \frac{p_2}{T_2}$ $\frac{P}{T}$ = a constant

Experiment 2: **Charles' law** $\frac{V_1}{T_1} = \frac{V_2}{T_2}$ $\frac{V}{T}$ = a constant

Experiment 3: **Boyle's law** $p_1V_1 = pV_2$ PV = a constant

TOP TIP

The mass of gas must remain constant.

Combining the equations gives:

$\frac{p_1V_1}{T_1} = \frac{p_2V_2}{T_2}$ $\frac{pV}{T}$ = a constant. This is the general gas equation.

TOP TIP

The temperature of gas must be in the Kelvin scale.

Example A balloon has a volume of 30 cm³. The temperature is 20°C and the pressure is $1{\cdot}01 \times 10^5$Pa. Assuming there is no loss of gas from the balloon, calculate the new volume of the balloon when the temperature drops to 7°C and the pressure drops to $1{\cdot}00 \times 10^5$Pa.

First step is to change temperature scale: 20°C becomes 293K, 7°C becomes 280K

$$\frac{p_1V_1}{T_1} = \frac{p_2V_2}{T_2} \qquad \frac{1{\cdot}01\times10^5\times30}{293} = \frac{1{\cdot}00\times10^5\times V_2}{280}$$

$293 \times 1{\cdot}00 \times 10^5\ V_2 = 280 \times 1{\cdot}01\times10^5 \times 30$ $V_2 = 29\ \text{cm}^3$.

The general equation can be used in all gas calculations instead of the three individual equations.

A constant quantity will simply cancel, e.g. $T_1 = T_2 \Rightarrow$ equation 3. So the general gas equation is the most useful one to remember – it's four equations in one!

Remember

In the pressure law, mass and volume are constant.

In Charles' law, mass and pressure are constant.

In Boyle's law, mass and temperature are constant.

In the general gas law, the mass of gas remains constant.

Quick Test 38

1. A deep-sea diver is down where the pressure is 3×10^5 Pa. He breathes out air bubbles of volume 2×10^{-6} m³. What volume will they have at the surface where the pressure is 1×10^5 Pa?
2. A 3 litre gas cylinder at 0°C has a pressure of 6×10^5 Pa. If the gas is used where the temperature is 27°C and the pressure is 1×10^5 Pa, calculate:

   (a) the volume the gas will occupy **(b)** the volume available.
3. What must be kept constant for all the gas equations?

Waves

Physics to learn	Longitudinal and transverse waves.
Revision guide	Describe waves, distinguish and give examples of longitudinal and transverse waves.

Energy waves

TOP TIP

Waves are used throughout the telecommunications, electrical and medical industries.

A wave is a movement of energy. Energy can be transferred from one place to another by a wave.
In some waves, energy moves through materials but other waves are simply a movement of energy though space.

There are many types of waves including:

water waves, sound waves, light, heat, radio and other electromagnetic waves.

Energy moves across the ocean and the water moves up and down.

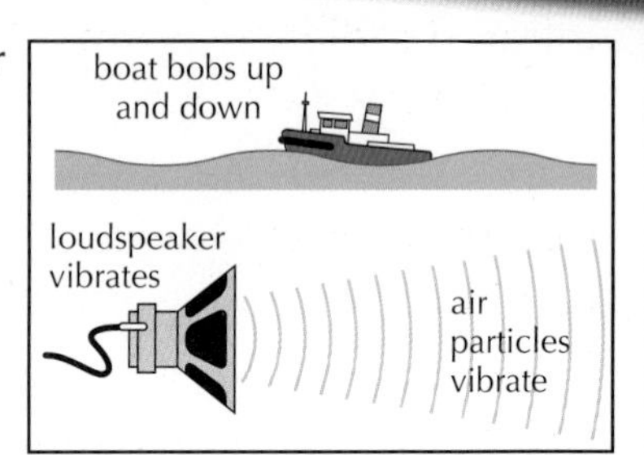

Sound energy travels through the air and the particles vibrate.

Radio waves can travel through space as well as through the air.

Longitudinal waves

Sound requires a medium or particles to travel through. Sound can pass through solids, liquids and gases but cannot pass through a vacuum. Sound is a **longitudinal** wave. The speaker cone below vibrates and the air particles vibrate in the same direction. The vibrations are in line with the direction of movement of energy.

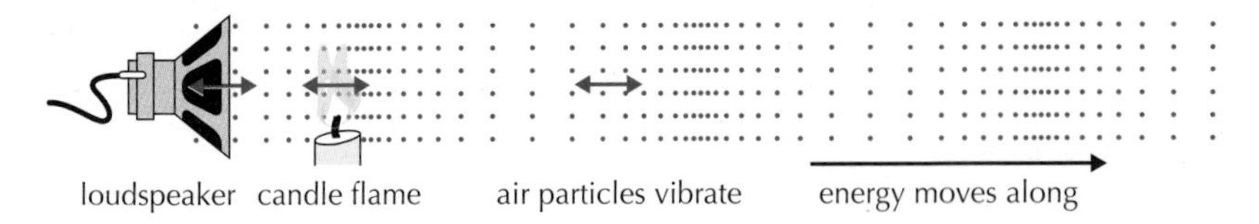

Wiggling a slinky can show that energy moves along but the coils just vibrate.

Wavelength
Amplitude
Rarefaction
Compression

When the particles are compressed this is labelled **compression** and when the particles are spaced out this is known as **rarefaction**.

The more energy that is put into a wave, the greater its **amplitude**, that is, the particles or molecules will vibrate more. Amplitude is measured in metres (m). Amplitude is the maximum distance a particle will be displaced from the normal position.

Transverse waves

Water waves and electromagnetic waves are all **transverse** waves. **Electromagnetic** waves include radio waves, microwaves, infrared radiation, visible light, ultraviolet radiation, X-rays and gamma rays.

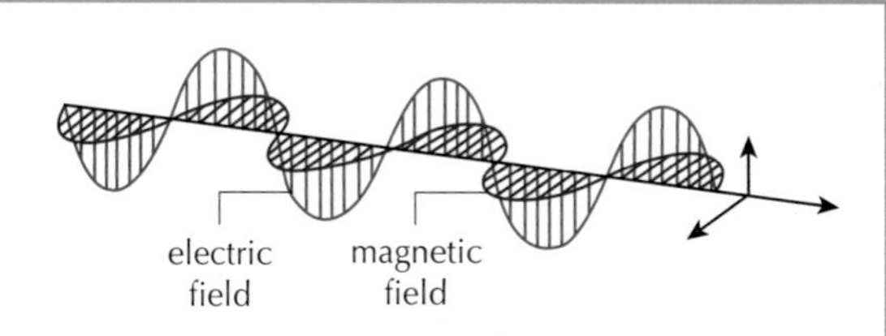

Water waves need the particles of water for the energy to travel through but electromagnetic waves are oscillations of electric and magnetic fields and so can travel through a vacuum.

The vibrations are at right angles to the direction of movement of energy. A wave can be sent along a rope to illustrate a transverse wave. The source moves up and down, the particles move up and down but the direction of energy movement is at right angles to this.

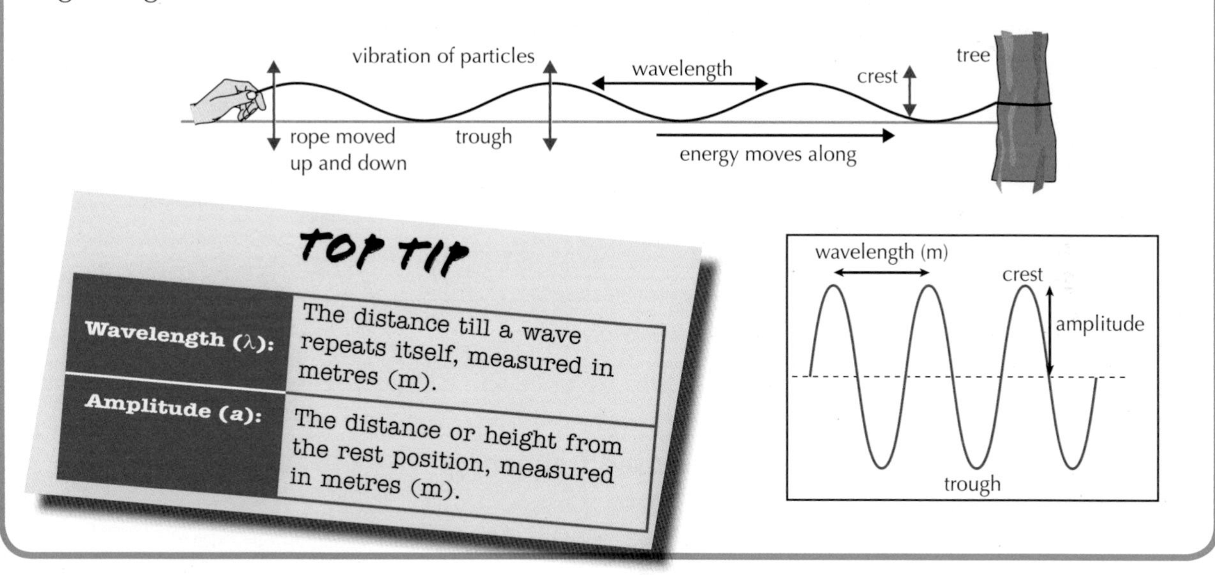

TOP TIP

Wavelength (λ):	The distance till a wave repeats itself, measured in metres (m).
Amplitude (a):	The distance or height from the rest position, measured in metres (m).

Quick Test 39

1. State what sound cannot travel through.
2. What is in a wave moving across the ocean?
3. What is the opposite of a compression?
4. How many waves are shown in the transverse rope above?
5. If the peak to trough height is 0·5 m, calculate the amplitude of the waves.
6. If the slinky on page 96 is 10 m long, calculate the wavelength of the waves shown.

Wave equations 1

Physics to learn	Wave measurements.
Revision guide	Use relationships between wave speed, frequency, period, wavelength, distance and time.

Waves

The signal generator creates electrical signals.

The loudspeaker turns electrical energy into sound energy, which is heard as longitudinal waves.

The oscilloscope changes electrical energy into light energy, which is seen as a transverse wave.

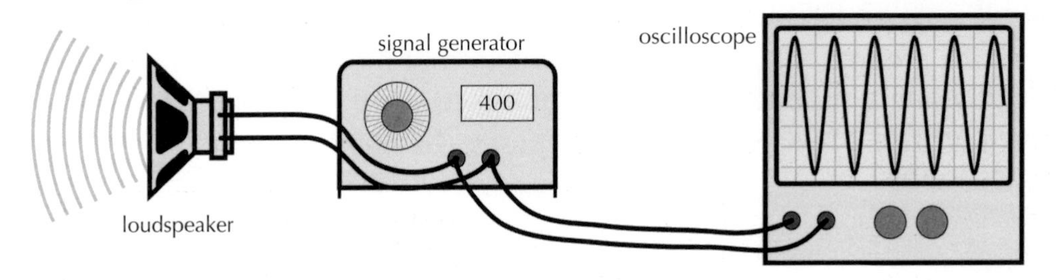

Adjust the **volume**: louder volume → **increased amplitude**.

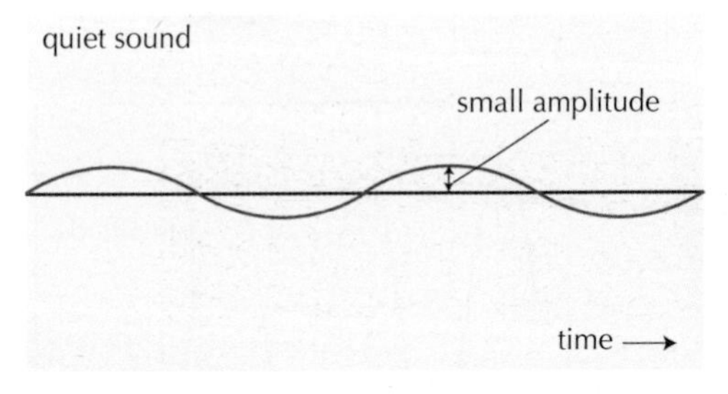

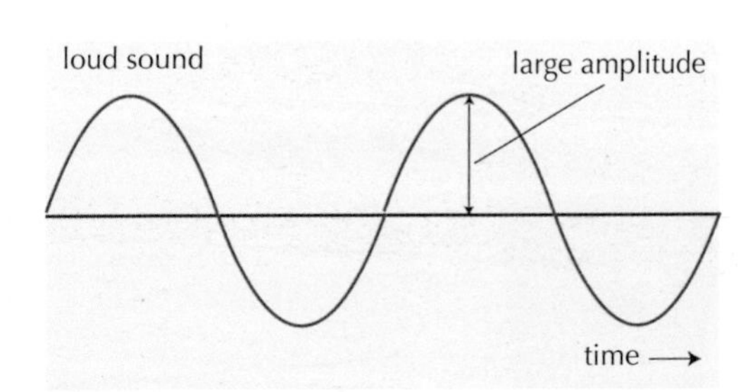

Adjust the **pitch**: higher pitch → **increased frequency**.

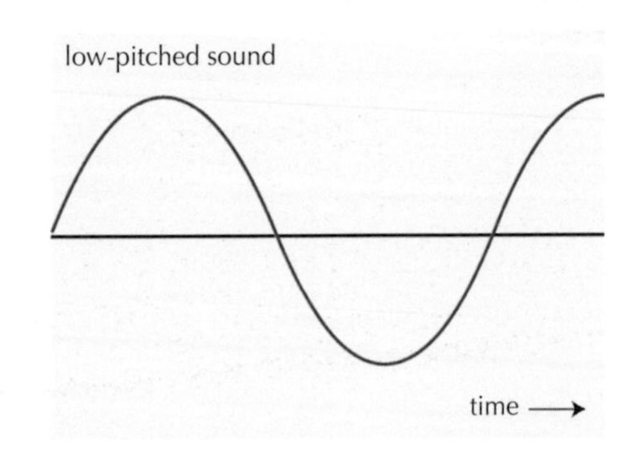

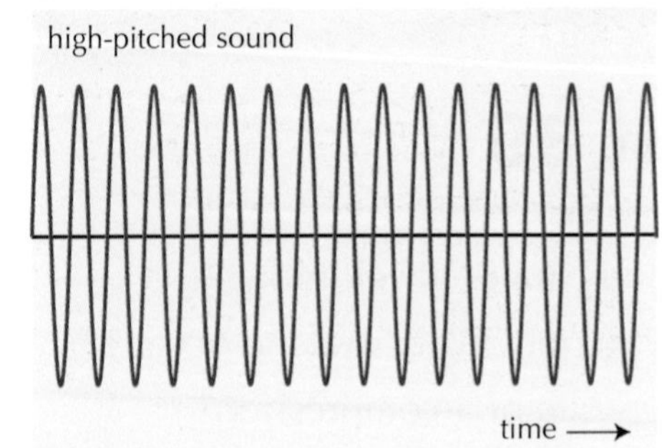

Period and frequency

The period (T) is the time to produce one wave and it is measured in seconds (s). If 10 waves take a time of 0·2 seconds, the time for one wave is $\frac{0 \cdot 2}{10} = 0{\cdot}02\,\text{s}$.

The frequency (f) of a signal is the number of waves produced in the unit time, e.g. 1 second.

$$frequency = \frac{number}{time} \qquad f = \frac{n}{t}$$

Frequency is measured in hertz, e.g. if 10 waves pass a point in a time of 0·2 seconds, the frequency, $f = \frac{n}{t}$ $f = \frac{10}{0 \cdot 2}$ $f = 50\,\text{Hz}$.

If we know the time for one wave, the period, then the equation for frequency becomes $f = \frac{1}{T}$. The equation for period is the opposite: Period, $T = \frac{1}{f}$.

Wavelength and frequency

Transverse water waves can be studied in a ripple tank.

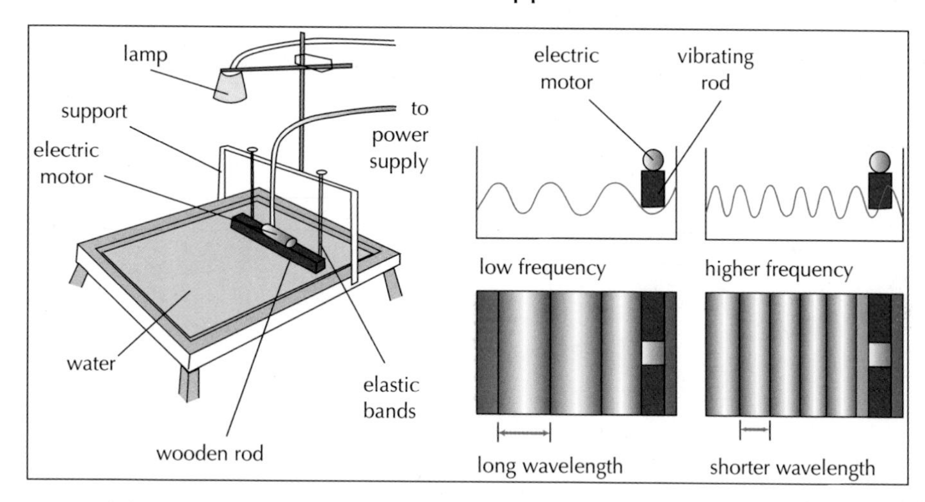

The frequency of the motor and rod is the same as the frequency of the water waves. Increase the frequency and the wavelength decreases.

The speed of the water waves does not change as the frequency and wavelength change, provided the waves move in the same depth of water.

$$speed = \frac{distance}{time} \qquad v = \frac{d}{t}$$

TOP TIP

When we study waves on the electromagnetic spectrum we find that as the frequency increases, the wavelength decreases.

Wave equations 2

Physics to learn	Wave velocity or speed.
Revision guide	Use two different relationships for wave speed.

Measuring wave speed

The speed (v) is the distance travelled in unit time. Speed is measured in metres/second (ms^{-1}).

We can measure the speed of sound in air in the lab.

- Measure a distance, say, 1 m.
- Place two microphones this distance apart attached to an electronic timer or computer and interface.
- The timer starts timing when the sharp sound of the hammer passes microphone 1 and stops timing when the sound passes microphone 2.
- Then calculate using $speed = \frac{distance}{time}$ $v = \frac{d}{t}$

The speed of sound is around 340 ms^{-1} in air. The time for sound to travel 1 m will be around 3 ms so the electronic timer will need to measure milliseconds or microseconds.

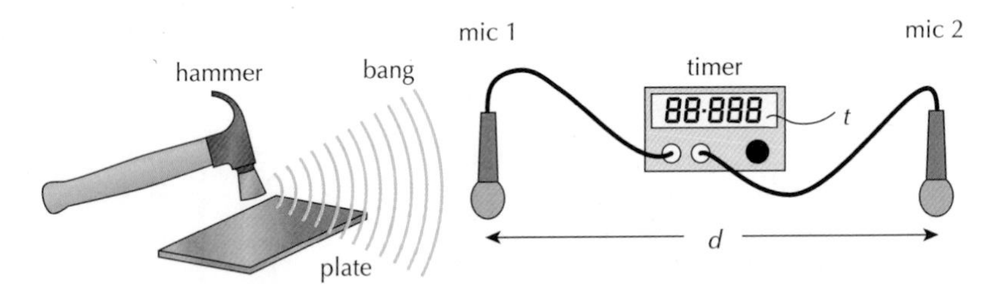

The speed of light is so fast that on Earth light appears to travel instantly. During thunder and lightning we do notice the large difference between the speed of sound and the speed of light.

The speed of light is accepted to be 3×10^8 ms^{-1} in a vacuum or in air. The speed of light is almost 1 million times faster than the speed of sound.

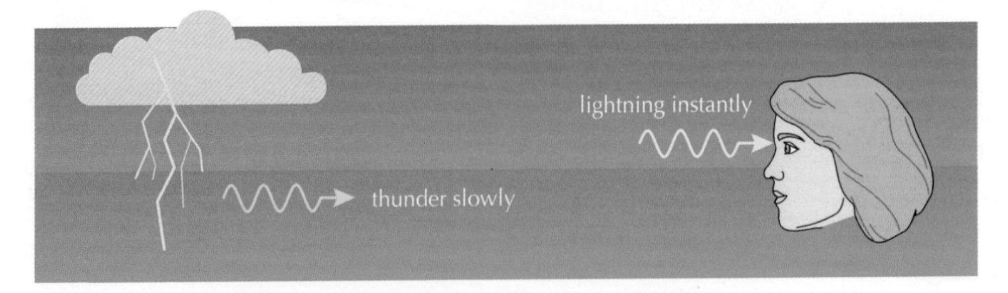

The wave equation

The speed of a wave is also related to its frequency and wavelength. If the distance travelled by a wave is one wavelength, the time taken is one period.

$v = \frac{d}{t}$ $v = \frac{1\lambda}{1T}$ Now $f = \frac{1}{T}$

so $v = f\lambda$ (the 'wave equation'.)

For example: a wave has a frequency of 15 Hz and has a wavelength of 2 cm. Calculate its speed. Remember to change cm to m.

$v = f\lambda = 15 \times 0{\cdot}02 = 0{\cdot}3\ ms^{-1}$

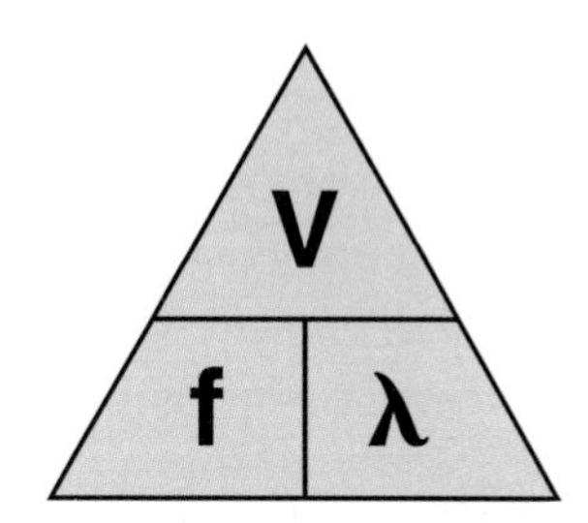

$v = f\lambda$ $\lambda = \frac{v}{f}$ $f = \frac{v}{\lambda}$

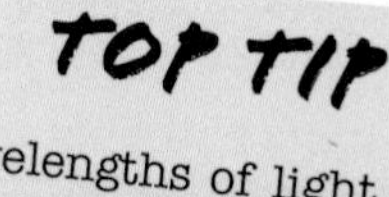

TOP TIP

The wavelengths of light are from 400 to 700 nanometers (nm).

TOP TIP

In a rearranged wave equation, v remains on the top.

Wave measurement summary

Wavelength (λ): The distance till a wave repeats itself.	metres (m)
Amplitude (s): The distance or height from rest.	metres (m)
Frequency (*f*): The number of waves in unit time.	hertz (Hz).
Period (*T*): The time for one wave (to pass).	seconds (s)
Velocity (*v*): The distance travelled in unit time.	metres/second (ms^{-1})
Velocity (*v*): The product of frequency and wavelength.	metres/second (ms^{-1})

Wave equation summary

$f = \frac{n}{t}$ $f = \frac{1}{T}$ $T = \frac{1}{f}$ $v = \frac{d}{t}$ $v = f\lambda$

Quick Test 40

1. A wave has a frequency of 10 Hz. What does this mean?
2. Thunder travels 1 km. Calculate approximately how long this takes.
3. There are 20 waves passing a point in 4 s. Calculate the:
   (a) frequency
   (b) period
4. Calculate the speed a wave of frequency 5000 Hz and length 0·02 m is travelling at.
5. What is the wavelength of sound of frequency 256 Hz (middle C) in air?

Diffraction

Physics to learn	Describe diffraction of waves.
Revision guide	Explain the effect of wavelength on sound, water, radio and microwaves.

Diffraction

Waves can bend around an object or obstacles. For example, the line of sight is blocked but sound is still heard because of the sound spreading out from a doorway. The waves do not simply pass straight through but spread out after passing through the gap.

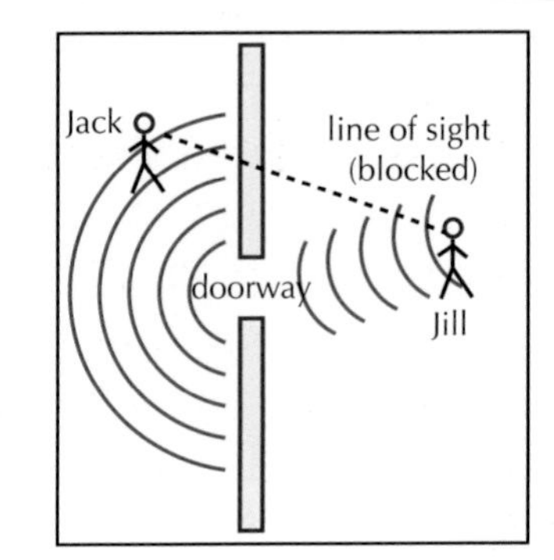

The bending, called **diffraction**, occurs at both edges or at gaps. We can think of a gap as being made of two edges. We can examine diffraction in a ripple tank.

Edges

As the waves pass the edge the energy spills to the side and the waves change from plane wavefronts to circular wavefronts.

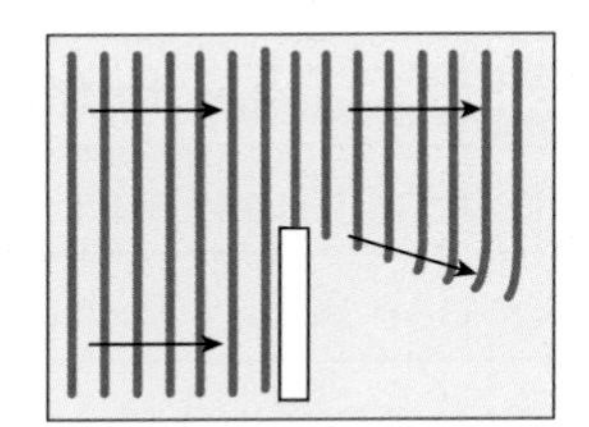

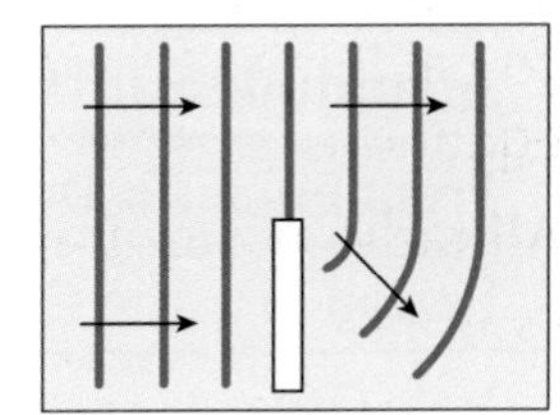

Long wave (LW), low frequency wavelengths diffract more than short wave (SW), high frequency wavelengths.

Gaps

A gap is created that is wider than the wavelength of the waves. Through the centre, the waves continue undisturbed. At each edge the waves show diffraction.

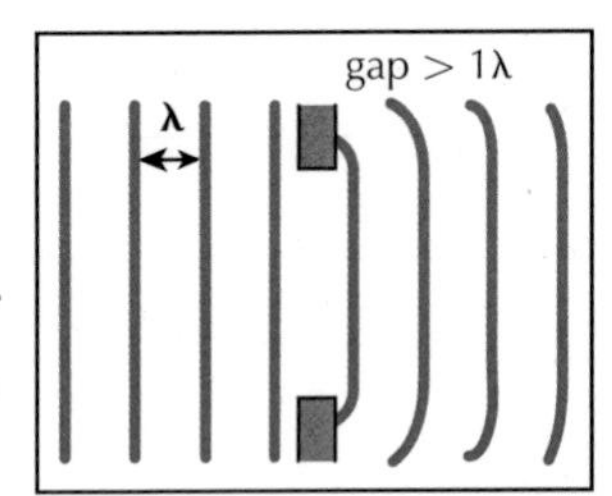

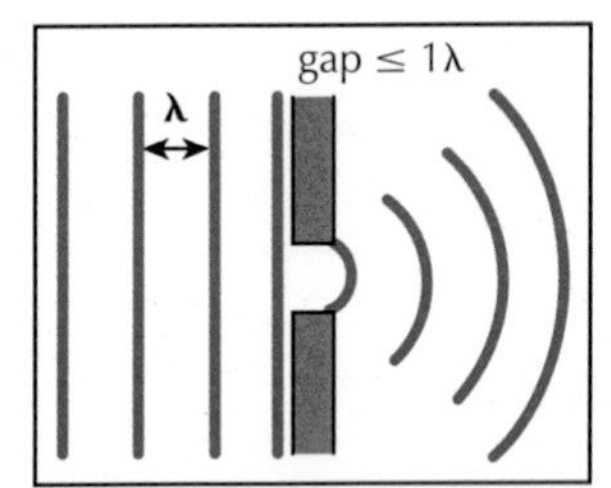

Circular wave-fronts are produced when the width of the gap is less than or equal to the size of the wavelength of the waves.

The wavelengths of sounds are about the same size as the width of doorways.

TOP TIP

If the wavelength decreases, the gap also needs to decrease to see diffraction.

Radio waves

Diffraction of radio

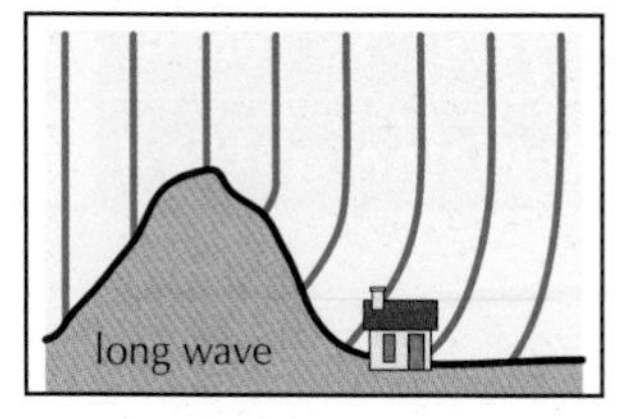

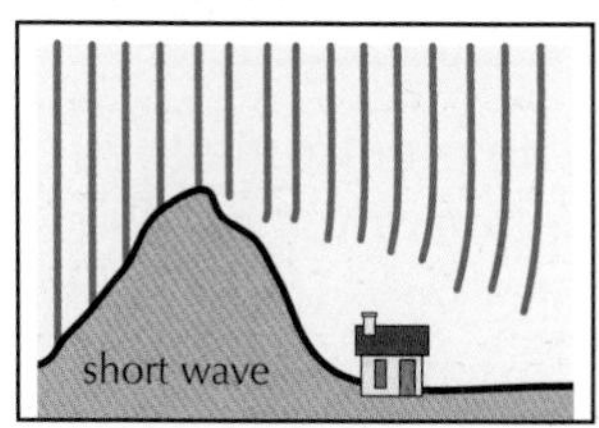

Radio waves have the longest wavelengths of the electromagnetic wave spectrum and range from about 100 kilometres to 1 millimetre.

The longest wavelengths will travel the furthest over hills because of their ability to diffract well.

Analogue TV waves are also radio waves but these have shorter wavelengths than radio station broadcast waves.

Digital satellite TV waves are also radio waves. These are known as microwaves as they occupy the shortest of the radio wavelengths.

Microwaves are used for satellite communication – they travel well in straight lines to and from the satellites in space because they exhibit little diffraction. Satellite receivers need line-of-sight views of the satellite in space.

TOP TIP

Radio waves have a longer wavelength than TV waves. Radio waves diffract more than TV waves.

Calculating wavelength

To calculate the wavelength of a radio wave we can use the wave equation.

All radio waves travel at 3×10^8 ms^{-1} through the air or space.

Example 1

What is the speed of radio waves whose wavelength is 1 millimetre?

You do not need to calculate this one! The speed of radio waves is $3 \times 10^8 ms^{-1}$.

Example 2

If the frequency of a radio wave is 2 GHz, what is its wavelength?

$$\lambda = \frac{v}{f} = \frac{3 \times 10^8}{2 \times 10^9} = 0{\cdot}15\text{m}$$

Example 3

Calculate the frequency of a radio wave of wavelength 2km.

$$f = \frac{v}{\lambda} = \frac{3 \times 10^8}{2 \times 10^3} = 1{\cdot}5 \times 10^5\text{Hz}$$

Quick Test 41

1. What is the meaning of diffraction?
2. What wavelengths bend most?
3. What are microwaves?
4. Why are radios still usable behind mountains?
5. Why does a satellite dish need to 'see' the satellite?

Electromagnetic spectrum: facts

Physics to learn	Bands of the electromagnetic spectrum.
Revision guide	Relative frequency and wavelength. Sources and applications.

Bands of the spectrum

The electromagnetic spectrum is made up of a very wide range of wavelengths. All the waves are transverse. The waves are put into bands or regions depending on how they act on matter or how they are made.

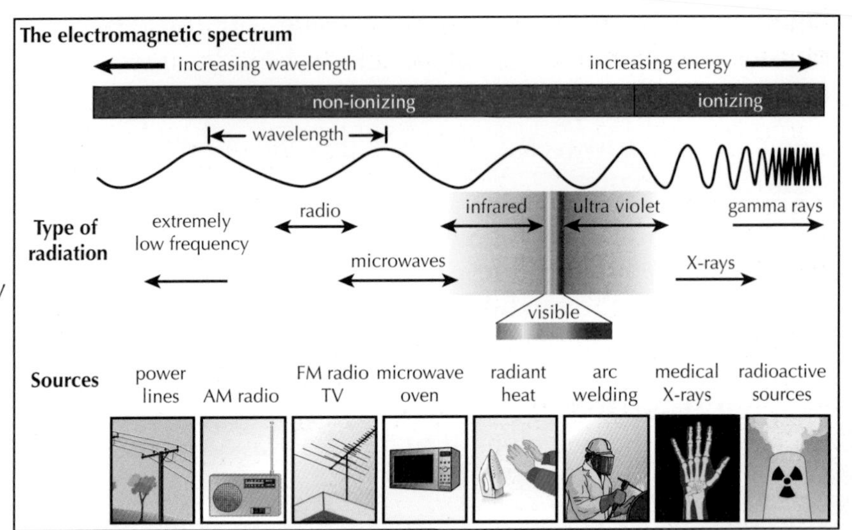

Sources and applications

Gamma rays

Gamma rays can be emitted from natural materials (such as some rocks, e.g. granite) and from man-made materials (such as some materials found in power stations). A lot of gamma rays are produced in the Universe and are travelling through space. Gamma rays have very high frequency, very short wavelength and high energy. They can be detected with photographic film or with a geiger counter. They can be used in radioactive tracers or to kill cancer cells. Instruments and syringes can be sterilised with gamma rays to kill bacteria. Gamma rays penetrate so well they can also be used to detect cracks in metals.

X-rays

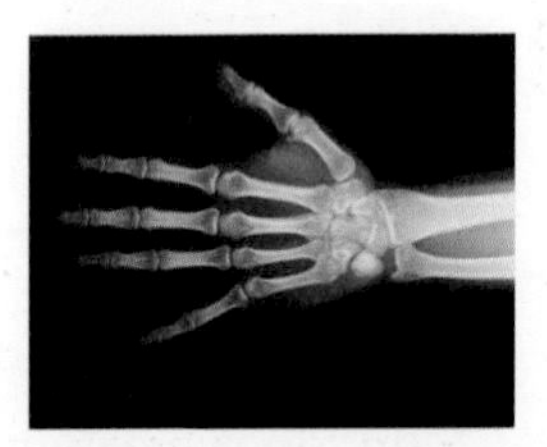

X-rays have high frequency and high energy. X-rays are produced from X-ray tubes. Hot gases in the universe also emit X-rays. They can penetrate tissue and be detected on photographic film. Doctors and dentists use X-rays to examine our bones and teeth.

Ultraviolet

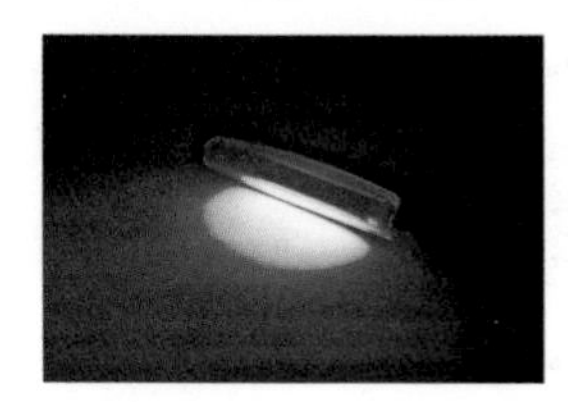

UV tubes and the Sun are sources of ultraviolet. They cause certain materials, often white, to fluoresce. Although the atmosphere stops many of these rays reaching us on Earth, they can cause suntans or give us skin cancer. Hidden marks on banknotes will fluoresce under a UV lamp. UV has a higher frequency and energy than visible light.

Visible light

Light is emitted from hot filaments of lamps, from stars and even fireflies. In physics, white is not a colour at all, but rather the combination of all the colours of the visible light spectrum. In 1672, when describing his discovery that light could be split into many colours by a prism, Isaac Newton gave the seven colours as red, orange, yellow, green, blue, indigo, violet. Violet has the shortest wavelength and red the longest.

Infrared

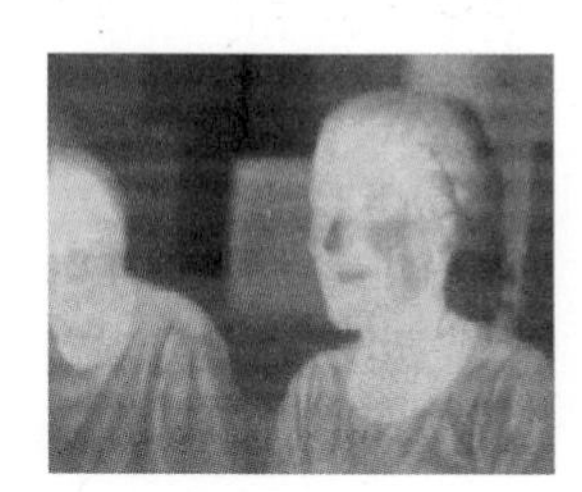

Infrared rays (heat waves) have a lower frequency than visible light rays. Infrared rays are absorbed by our skin and we feel warm. Invisible heat rays are given out by all warm bodies. Thermograms are colour heat photos of this radiation. If an infrared camera takes your photo you will not notice it as the rays are invisible. Burglars beware!

Microwaves

Above radio we find microwaves with a higher frequency (in the GHz range). Microwaves diffract (bend) very little though, compared with radio waves. Microwaves are sent to and from satellites. Microwaves are used from space by astronomers to find out about the structure of galaxies. Microwaves are used with mobile phones and for cooking.

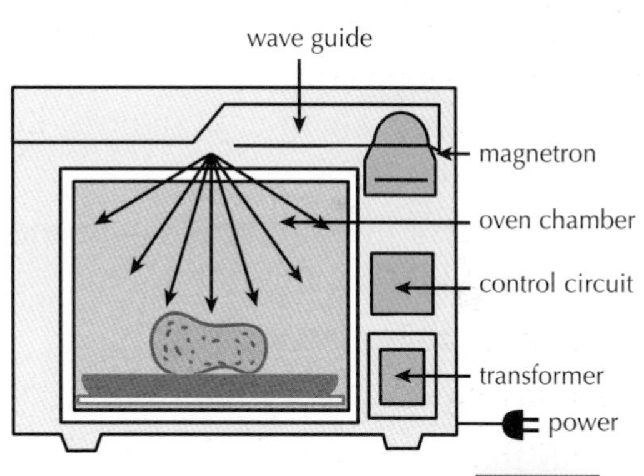

TV and radio

Radio and TV waves have the longest wavelengths in the electromagnetic spectrum. They are transmitted for communication but are also emitted by stars and gases in space.

Quick Test 42

1. List the bands of the electromagnetic spectrum in order.
2. What bands have a shorter wavelength than light?
3. What bands have a longer wavelength than light?
4. Find out when the electromagnetic spectrum was created.
5. List sources of electromagnetic waves.
6. List applications of electromagnetic waves.

Electromagnetic spectrum: data

Physics to learn	Relationships: frequency and energy, frequency and wavelength.
Revision guide	Velocity calculations with frequency and wavelength.

Velocity

The electromagnetic spectrum has a range which varies from extremely short to extremely long waves.

Most of the electromagnetic spectrum is invisible to our eyes. Only the colours of light are visible. It is known that charge has an electric field around it and that moving charges (such as in an electric current) give a magnetic field. In 1861, James Clark Maxwell made the connection between electricity, magnetism and the speed of light. He concluded that light was an electromagnetic wave.

Experimental measurements of the speed of light and electromagnetic waves can be made to a very high degree of certainty. More than this, by using knowledge of electricity and magnetism, physicists have calculated exactly what the speed of light and all electromagnetic waves should be: 299 792 458 ms^{-1} in a vacuum.

Velocity through a vacuum or air

For all our calculations we will use the speed of light and all electromagnetic waves to three significant figures, i.e. 300 million ms^{-1}.

The speed of light and all electromagnetic waves is 300 000 000 ms^{-1}.

Using scientific notation:

$$v = 3 \times 10^{8}\ ms^{-1}$$

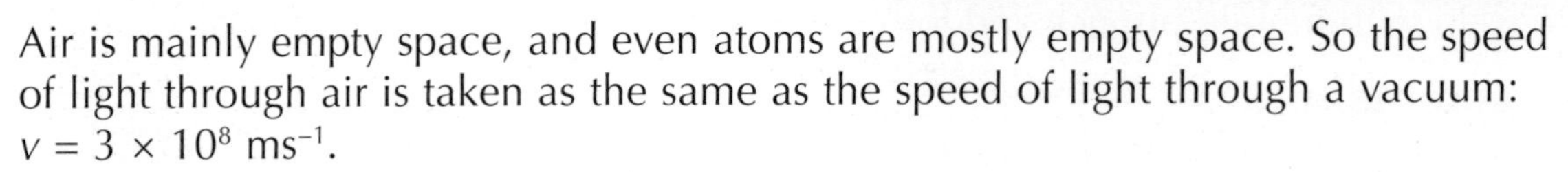

Air is mainly empty space, and even atoms are mostly empty space. So the speed of light through air is taken as the same as the speed of light through a vacuum: $v = 3 \times 10^{8}\ ms^{-1}$.

Velocity through dense materials

However, light does travel slower through optically dense materials. In normal glass, light travels at:

$$v_{\text{light in glass}} = 200\ 000\ 000\ ms^{-1}$$

or $$v_{\text{light in glass}} = 2 \times 10^{8}\ ms^{-1}$$

Frequency and wavelength

As we move through the electromagnetic spectrum from radio waves to gamma rays, the wavelength decreases dramatically. At the same time, and as the velocity does not change, this means that frequency increases dramatically.

As frequency increases and wavelength decreases the product has been found to be a constant and equal to the velocity.

Frequency, wavelength and velocity are related by the wave equation:

velocity = frequency × wavelength

$$v = f\lambda$$

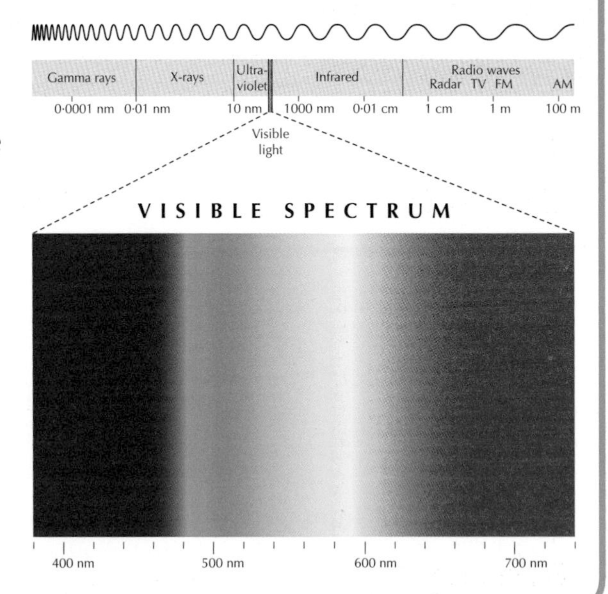

Frequency and energy

Two famous physicists, Max Plank and Albert Einstein, discovered that as frequency increases, the electromagnetic radiations have more energy. They considered the radiations to be made of small bundles of energy called photons. Gamma ray photons have the most energy with X-rays photons next.

High frequency gamma rays and X-rays have good use in medicine and industry due to their energy and penetration power. Ultraviolet also can damage our skin but we call this a suntan!

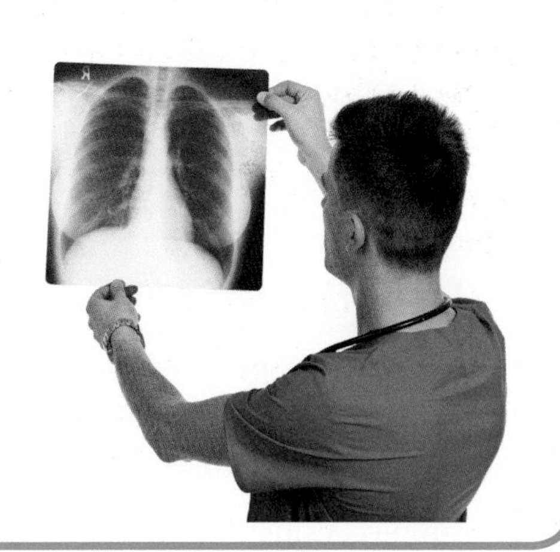

Quick Test 43

1. What is the speed of radio waves in a vacuum?
2. What is the speed of X-rays through the air?
3. What is the speed of light through normal glass?
4. A gamma ray has a frequency of 1×10^{12} Hz. Calculate its wavelength.
5. A radio wave has a wavelength of 1 km. Calculate its frequency.
6. An electromagnetic wave has a length of 3μm. What part of the spectrum does it belong to?

Reflection and Refraction

Physics to learn	Reflection of light. Refraction of light.
Revision guide	Velocity, angles and applications.

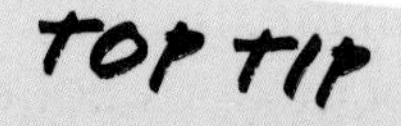

Angles are always measured between the ray and the normal. A normal is a line at 90° to the surface, which is labelled 0°.

Reflection from a plane mirror

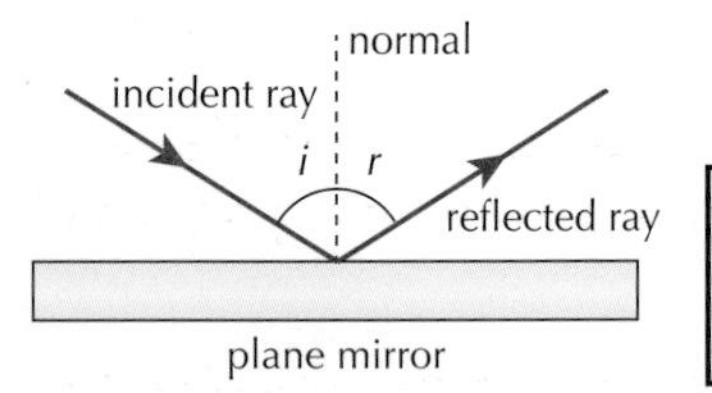

angle of incidence = angle of reflection

$$i = r$$

When a ray of light strikes a plane mirror it is reflected so that the angle of incidence is equal to the angle of reflection.

Refraction and velocity

Refraction occurs when light moves from one optical density of material to another.

Glass is optically more dense than air. If light moves from air to glass, velocity decreases. Light travels at 3×10^8 ms^{-1} in air and about 2×10^8 ms^{-1} in glass.

If we could observe the wavefronts or the 'crests' of light we would notice that they behave in a similar manner to water waves when they cross a shallow region in a ripple tank. As velocity decreases, wavelength decreases and the ratio of $\frac{v}{\lambda}$ remains constant.

Frequency is constant, $\left(f = \frac{v_{air}}{\lambda_{air}} = \frac{v_{glass}}{\lambda_{glass}} \right)$

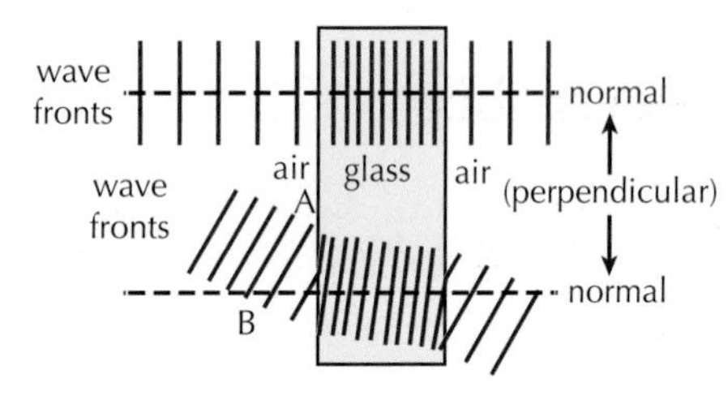

refraction through a parallel plate

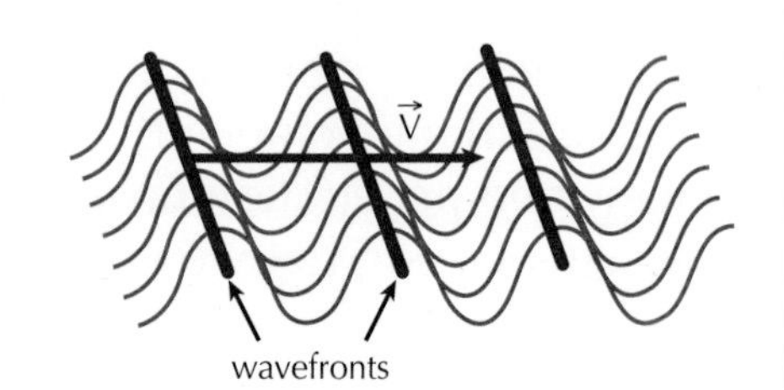

Refraction and direction

The change in velocity during refraction at an angle will produce a change in direction.

- **Air to glass:** When a ray of light enters a more dense medium it slows down and bends towards the normal.
- **Glass to air:** When a ray of light enters a less dense medium it speeds up and bends away from the normal.

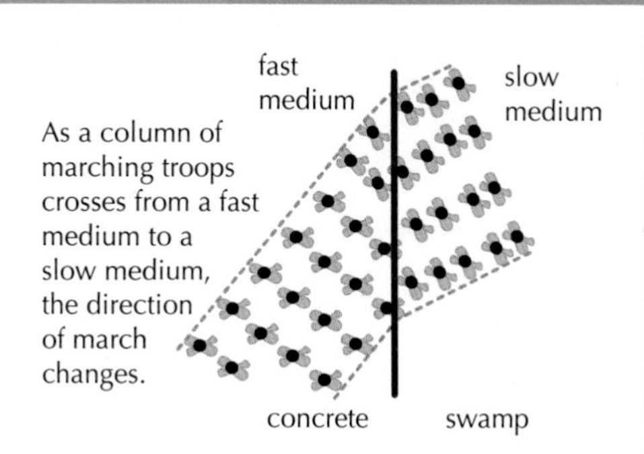

The angle in the more dense medium is smaller than the angle in the less dense medium. This rule is independent of whether the ray is entering or emerging from the medium.

If we have an eye defect, refraction allows additional lenses to further change the direction of light rays, thus correcting eye defects.

A **convex** converging lens.

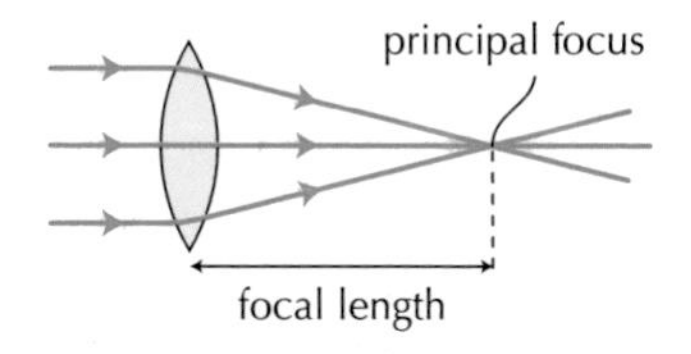

A **concave** diverging lens.

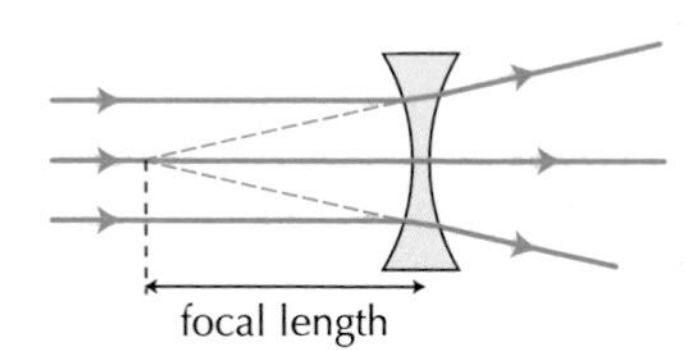

Reversibility of light

Light rays are reversible. A ray enters glass at 60°, it will refract to 35°. In reverse, the ray meeting the boundary at 35° leaves the glass at 60°.

Air to glass

60°
air
glass
35°

Glass to air

60°
air
glass
35°

Quick Test 44

1. State what changes occur when light goes straight into a tank of water.
2. What happens to the frequency of waves during refraction?
3. What happens to the wavelength of waves during refraction?
4. What direction does light take going from glass to air at an angle?
5. Define angle of incidence.

Angles of refraction

Physics to learn	Knowledge of refraction. Identification of angles.
Revision guide	Use diagrams and follow paths for refraction.

Refraction

Refraction occurs when waves pass from one medium to another.

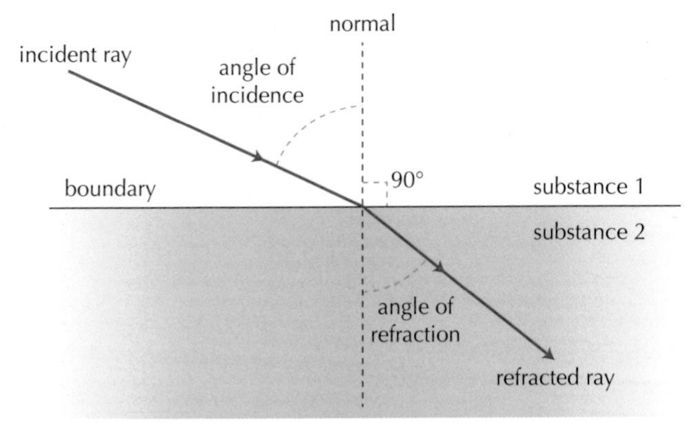

When refraction occurs there is a change of wave speed, a change in wavelength but no change in frequency of the waves. Note there is no change in direction when the angle of incidence is 0°.

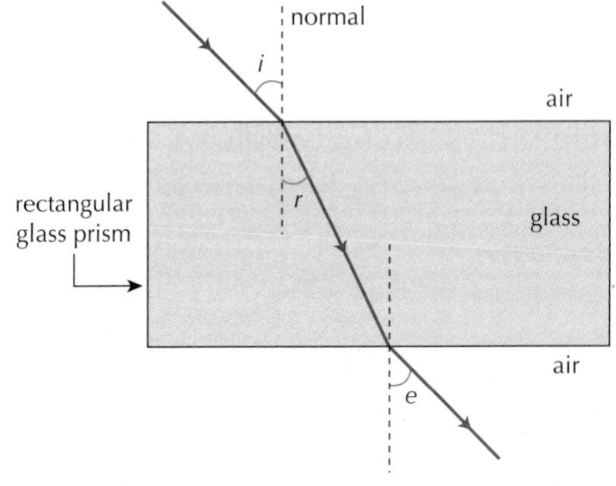

Both wave speed and wave length will decrease when entering a more dense medium. Both wave speed and wave length will increase when entering a less dense medium.

The colours or frequencies of white light can be separated by refraction.

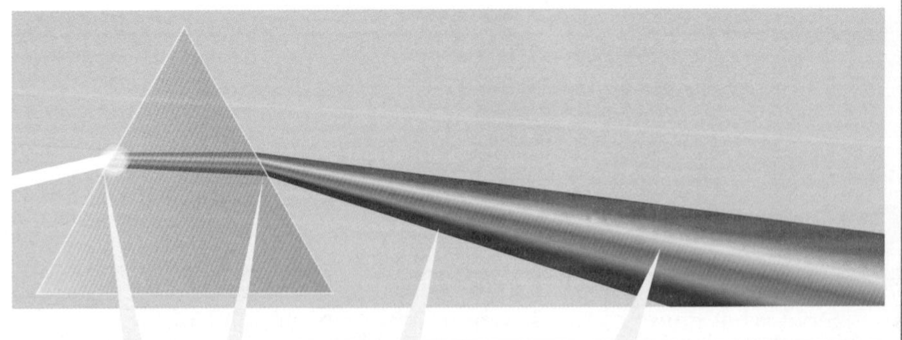

Light passing through a prism is refracted twice- once when it enters and once when it exits.

Violet light, which has a short wavelength, is refracted more than red light, which has a long wavelength.

You can see the colors of the rainbow when white light is separated by a prism.

Limits of refraction

A ray of light is directed into a glass prism. We know that when light enters a more optically dense material it refracts and the angle in the glass is smaller than the angle in the air.

The incident ray makes an angle of incidence, *i*, with the normal.

The refracted ray makes an angle of refraction, *r*, with the normal.

From air to glass: $r < i$

An experiment can be done to confirm this theory. As the angle of incidence increases, the angle of refraction increases also, but it is always smaller in the glass.

Angle of incidence, *i* (°)	Angle of refraction, *r* (°)
0	0
10	6·6
20	13·2
30	19·5
40	25·4
50	30·7
60	35·3
70	38·8
80	41·0
90	42·0

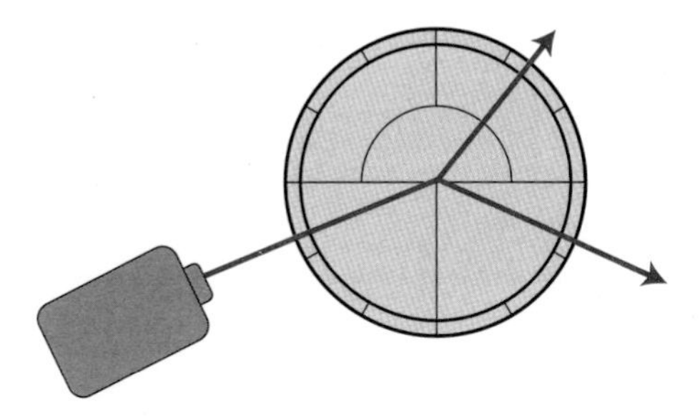

As the angle of incidence increases towards 90°, the angle of refraction is seen to reach a limit. In fact, if we try to illuminate the angles beyond this limit we find we cannot do so.

TOP TIP

The limit angle is what we know as the critical angle.

Quick Test 45

1. State what is meant by the critical angle.
2. State what is meant by total internal reflection.
3. How do you measure the critical angle?

Nuclear nature

Physics to learn	Nature of nuclear radiation.
Revision guide	Knowledge of the structure of the atom, and the nature of alpha, beta and gamma radiation.

The atom

Atoms are the building blocks of elements and each element is made of only one unique type of atom.

Atoms have subatomic particles and all atoms have a nucleus at the centre with tiny electrons in orbit around this.

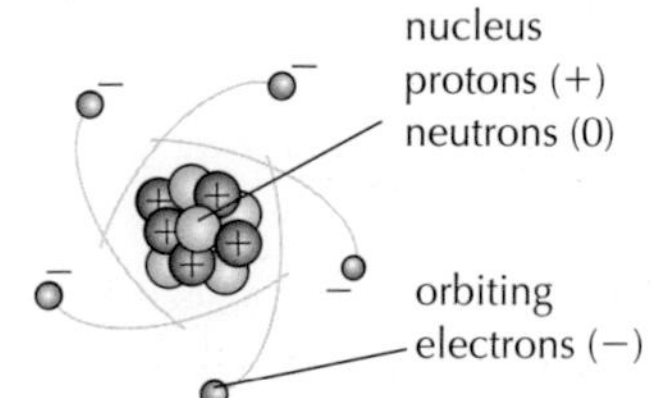

Nuclear radiation emits from certain nuclei and all the types of radiation we study here are called 'nuclear radiation'.

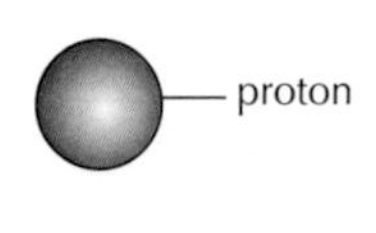

hydrogen nucleus

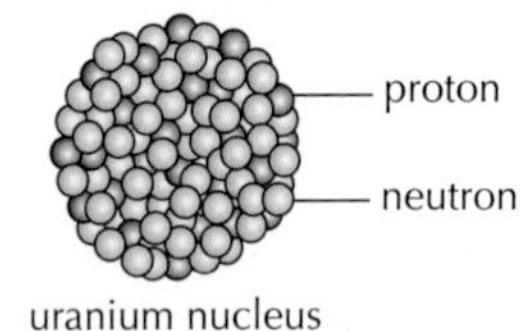

uranium nucleus

TOP TIP

Atoms are neutral if the number of electrons = the number of protons.

The nucleus is made of neutrons and protons. Over 99% of the mass of an atom is in the nucleus.

Electrons are not part of the nucleus. They orbit the nucleus at certain levels with huge space between themselves and the nucleus.

The nature of radiation

There are three types of radiation emitted from the nucleus that we call nuclear radiation: alpha particles (α), beta particles (β), gamma rays (γ).

Alpha particles (α)

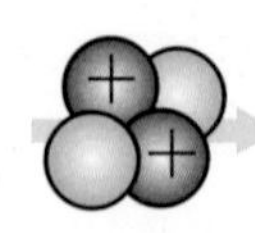

An alpha particle is made of two protons and two neutrons emitted from a large unstable nucleus. Alpha has two protons, therefore a positive charge of +2.

The mass of protons and neutrons is said to be 1 atomic mass unit or 1 amu each. This means the mass of an alpha particle is 4 amu. That is the heaviest of our three nuclear radiations.

Alpha particles are ejected from the nucleus at very high speeds: about 5–10% of the speed of light. However, they are the slowest of our three nuclear radiations.

Beta particles (β)

A beta particle is a very fast electron.

Beta particles are emitted from the nucleus when a neutron breaks up into a proton and the fast electron.

$$ {}^{1}_{0}n \rightarrow {}^{1}_{1}p + {}^{0}_{-1}\beta $$

That electron is the beta particle. It has a lot of energy. Beta is a lot smaller than alpha. The mass of a beta particle is about 2000 times smaller than 1 amu or a neutron or proton. The beta particle has a charge of 1 and is negative, the same mass and charge as an orbiting electron.

The velocity of an ejected beta particle is faster than an orbiting electron.

Gamma rays (γ)

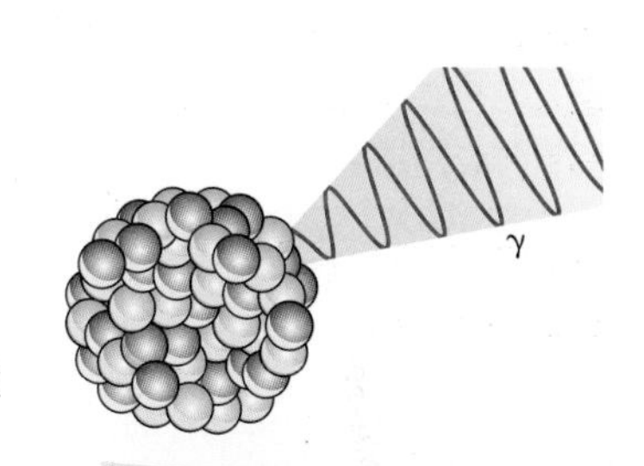

Gamma rays are bursts of energy that a nucleus emits to become more stable. On the electromagnetic spectrum these energy waves have a very high frequency and very short wavelength. Gamma rays have no mass or charge. Gamma rays travel at the speed of light. The speed of light is sometimes given as the symbol c.

Velocity of gamma rays, $v_{gamma} = 3 \times 10^8$ ms^{-1}.

As well as being emitted from nuclear radiation on Earth, our sky is full of bursts of gamma radiation from the hottest and most energetic objects in the Universe. Our atmosphere absorbs gamma rays to protect us.

TOP TIP

Remember all these radiations are nuclear. Alpha and beta are particles; gamma is a burst or wave of energy.

TOP TIP

Remember the speed of light, $c = 3 \times 10^8$ ms^{-1}.

Quick Test 46

1. State why an atom is normally neutral.
2. Where are the protons in an atom?
3. Which radiation is not a particle?
4. What is the charge on a neutron?
5. What charge is on an alpha particle?
6. What is the mass of a beta particle?
7. What does a gamma ray have?

Ionisation

Physics to learn	Ionisation.
Revision guide	Knowledge of ionisation and detection methods.

Ionisation

Atoms are normally electrically neutral. Ionisation occurs when an atom gains or loses one or more electrons.

Adding an electron to an atom creates a negative ion with more electrons than protons.

Removing an electron from an atom creates a positive ion with fewer electrons than protons.

Ionisation can be caused by radioactivity. Alpha (α), beta (β), and gamma (γ) are ionising radiations as they can ionise atoms they hit. When ionising radiations collide with atoms they can knock an electron away, leaving a positive ion.

Alpha particles produce a much greater ionisation than beta particles or gamma rays. Alpha particles cause the most ionisation because they are the largest and because they have the greatest charge of the radiations. Alpha particles give up their energy quickest and are absorbed in the shortest distance.

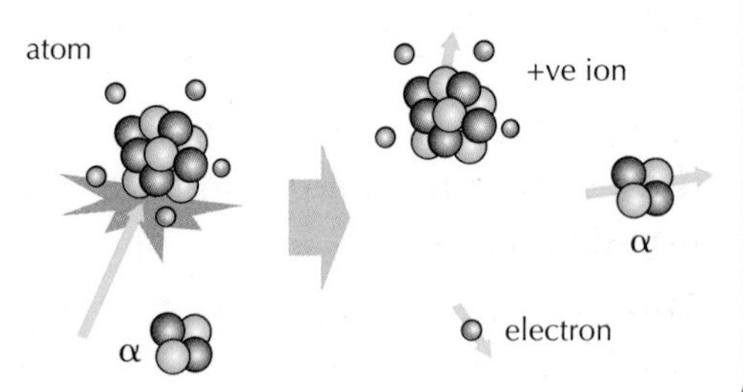

Penetration of ionising radiation

Radiation energy may be absorbed in the medium through which it passes. The absorption properties can identify the type of radiation.

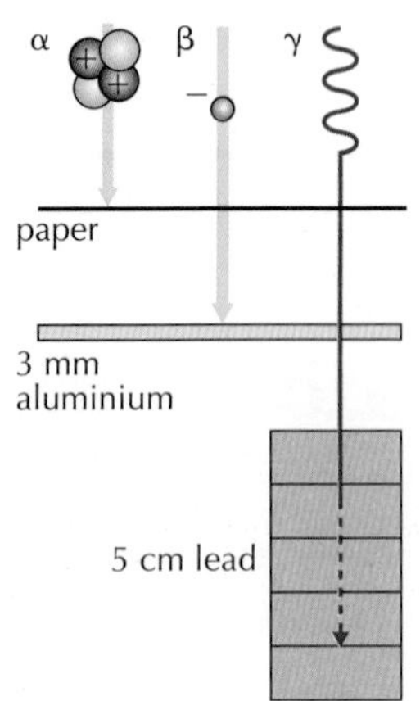

A slow alpha particle cannot penetrate more than about 5 cm of air. Alpha is easily stopped by a few sheets of paper and cannot penetrate your skin. (Swallowing an alpha emitter is lethal as they don't escape the body.)

Fast beta particles can penetrate through several metres of air before losing their energy. Beta particles are stopped by a few millimetres of aluminium.

Gamma rays can penetrate the Earth's atmosphere. Air does not absorb gamma. They have very high energies. They can only be stopped by several centimetres of lead or several metres of concrete.

Alpha causes the most ionisation so has the least penetrating power.

Gamma causes the least ionisation so has the highest penetrating power.

Identifying radiation

Alpha, beta and gamma radiations can also be identified by how they behave in magnetic or electric fields.

Deflection by magnetic field

In a magnetic field, charged particles are deflected. The alpha particles with a positive charge move in the opposite direction to the beta particles with a negative charge.

The mass of the particles also has an effect on how far they are deflected.

Alpha particles have much more mass than beta and this makes them harder to deflect in a magnetic field.

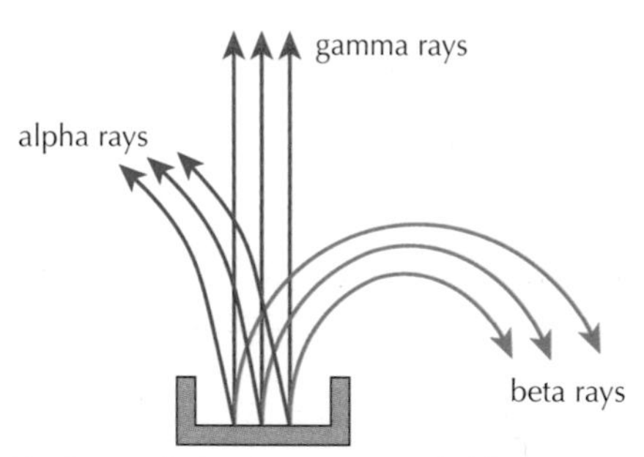

Nuclear radiation in a magnetic field.

Deflection by electric field

In an electric field, charged particles are also deflected. The alpha particles with a positive charge move in the opposite direction to the beta particles with a negative charge.

The mass of the particles also has an effect on how far they are deflected. Alpha particles have much more mass than beta and this makes them harder to deflect in an electric field.

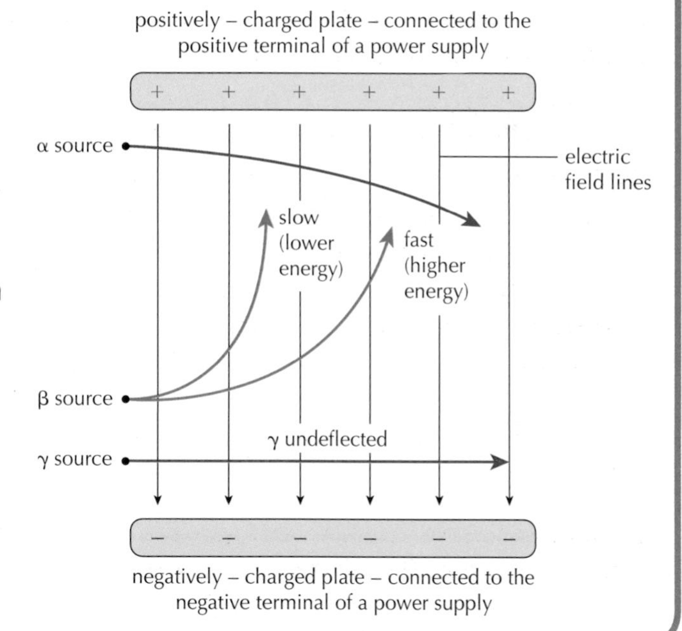

Quick Test 47

1. State what radioactivity does to atoms.
2. State which type of radioactivity is the most penetrating radiation.
3. State which type of radioactivity causes the greatest ionisation.
4. What happens to radiation energy as it passes through a material?
5. Name methods of distinguishing between the three types of radioactivity.

Nuclear activity

Physics to learn	Activity and background radiation.
Revision guide	Define and use relationships involving activity.

Activity

Radioactive decay takes place in a random manner. A radioactive substance contains many nuclei. We cannot say when any individual nucleus will decay, as the decay takes place at a random time, but with such a large number of nuclei in any sample we can observe the average number that are decaying in a certain time.

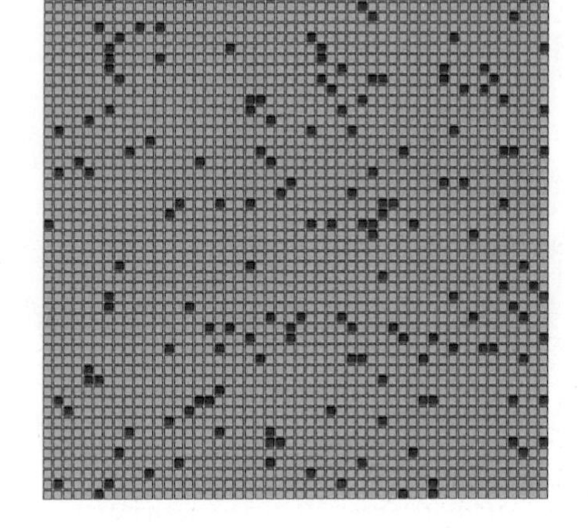

This rate of decay is known as the activity of the substance:

$$activity = \frac{number\ of\ nuclei\ decaying}{time} \qquad A = \frac{N}{t}$$

Activity of radioactive sources is defined as the number of nuclear decays per second, and is measured in becquerels (Bq). One becquerel is one decay per second.

Example

A sample of 1 g of uranium-238 has 720 000 decays in 1 minute. Calculate the activity of this source.

$$A = \frac{N}{t} = \frac{720000}{1 \times 60} = 12\,000\,\text{Bq} = 12\,\text{kBq}$$

Background radiation

Background radiation is all around us and has to be deducted from any measurements of radiation sources. We can count the radiation in becquerel (Bq), or measure the danger in sieverts (Sv). Background radiation comes from two types of source: natural and artificial.

Most background radiation from natural sources comes from the radioactive gases around us.

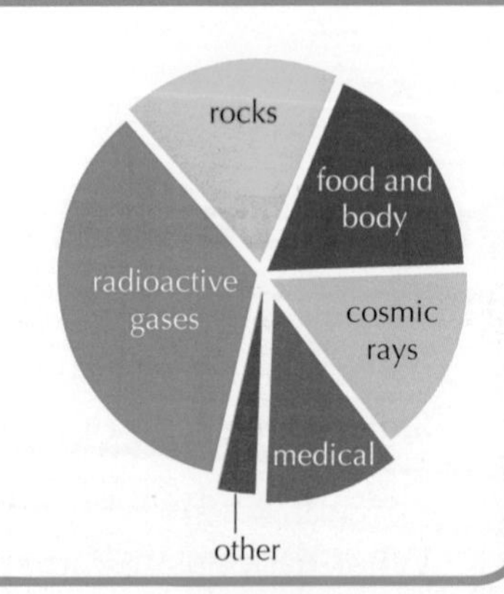

Natural source	Annual equivalent source (μSv)	Annual equivalent source (mSv)
Radioactive gases in air and buildings (radon and thoron)	800	0·8
Rocks of the Earth	400	0·4
In food and in our body	370	0·37
Cosmic rays from space	300	0·3
Total natural sources	**1870**	**1·87**

Most background radiation from artificial sources comes from the medical use of radiation.

Man-made source	Annual equivalent source (μSv)	Annual equivalent source (mSv)
Medical uses (X-rays)	250	0·25
Weapons testing	10	0·01
Nuclear industry (waste)	2	0·002
Other (job, TV, flight)	18	0·018
Total man-made sources	**280**	**0·28**

In addition to background radiation, annual effective dose limits have been set for extra exposure to radiation for the general public, and higher limits for radiation workers.

Extra safety limits:

- For the public: add 1 mSv yr^{-1}.
- For radiation workers: add 20 mSv yr^{-1}.

TOP TIP

The background radiation adds up to about 2·2 mSv per year per person.

Quick Test 48

1. A sample of uranium has $1{\cdot}2 \times 10^6$ decays in 1 minute. Calculate its activity.
2. Another sample of 1g of uranium has an activity of 3·0 MBq. How many nuclei decay in 1 minute?
3. What is the effective dose a person will receive from background radiation in a year?

Nuclear applications

Physics to learn	Applications of nuclear radiations.
Revision guide	Awareness of safety and industrial and medical applications.

Safety first

The early discoverers of radioactivity were not aware of its danger. Henri Becquerel first discovered uranium rays in 1896. At the same time, the husband and wife team of Marie and Pierre Curie were researching in Paris. Marie invented the word 'radioactivity' and discovered the radioactive elements polonium and radium in 1896–1900.

Radioactive elements are naturally occurring and they can be used in science and medicine. Too much radiation causes tissue damage, cancer, genetic disorders, and even death. Marie Curie died as a result of her research into and ongoing exposure to radioactivity.

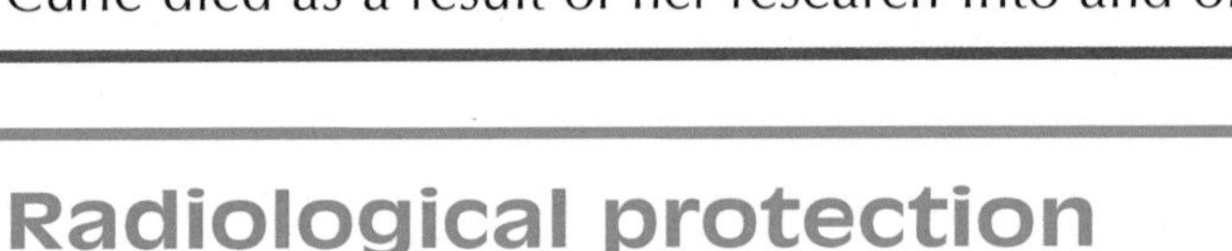

Radiological protection

Exposure to radiation can be harmful. The equivalent dose is reduced by shielding, by limiting the time of exposure or by increasing the distance from a source.

1. **Shielding:** Lead and concrete are good absorbers of radiation. A radioactive source is often kept in small lead cases. A radiographer will wear a lead-lined apron. Schools or hospitals will often keep their sources behind brick or concrete walls. A nuclear plant will make use of thick concrete round its reactor.
2. **Limiting exposure time:** Sources should be brought in and used in as short a time as possible.
3. **Increasing the distance from the source:** Radiation often spreads out like the rays of a light bulb. This means that intensity decreases rapidly with distance. Use tongs to increase the distance when in use. Just putting a source on the other side of the room when awaiting use may dramatically reduce exposure.

Medical applications

A radioactive tracer can be injected into a patient and this collects on organs or bones where there is a problem. This allows diagnosis. The radiations emitted travel out of the body to be examined with a gamma camera. Care is taken to use a radio-isotope with a short half-life and a suitable dose.

When ionising radiation is used to treat cancer, the radiotherapy treatment is prescribed in units of Gy. When radiopharmaceuticals are used, they will usually be given in units of becquerel. Health risks are discussed in the sievert unit. Different units are used because they measure different things.

Radiotherapy is used to for the treatment of patients with cancer. Malignant cells can be given a high dose and the energy kills them. A source may be rotated around the patient. Any normal cells which get damaged can repair themselves.

low dose here causing no damage to healthy cells

high enough dose to kill cancer cells

γ source

Medical instruments can be sterilised with radiation. A strong gamma source destroys bacteria so they don't cause infection.

Industrial applications

Radiation workers need their exposure to radiation monitored. A film badge consists of photographic paper with different types and thicknesses of absorbing material such as plastic, aluminium and lead placed in "windows" in front of it. The penetration by nuclear radiation will blacken the photographic film. The type and amount of radiation can be determined.

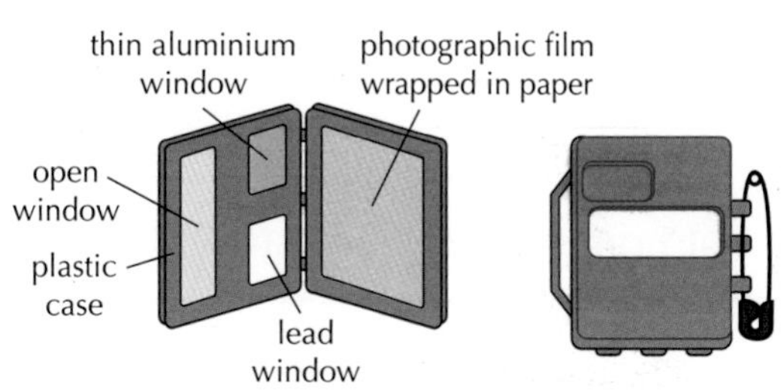

Radiographic testing involves penetrating radiation passing through a solid object, such as a weld, onto photographic film. A crack in the object will allow more radiation through and blacken the film more. An advantage of this method is that it uses non-destructive testing. This method can also be used on some pressure vessels and storage containers.

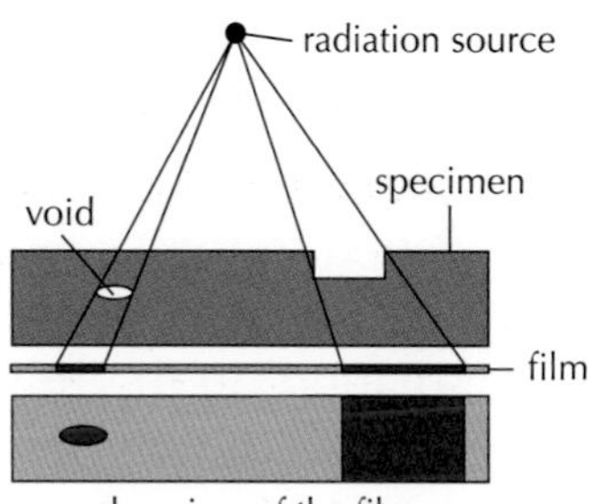

plan view of the film darker areas (after processing)

Quick Test 49

1. Why do radio-isotopes emit gamma rays?
2. Why is a source rotated during radiotherapy?
3. Where is the object placed when monitoring nuclear radiation?

TOP TIP

Further applications of radiation are found in the Nuclear power section.

Nuclear relationships

Physics to learn	Absorbed dose, Equivalent dose and Equivalent dose rate.
Success guide	Understand and use nuclear dose relationships.

Absorbed dose

When ionising radiation is absorbed by the human body, physicists can measure the energy put down by the absorbed particles and the mass of matter absorbing the radiation. From these, we can calculate the absorbed dose, which is the energy deposited per unit mass into the absorbing material.

$$absorbed\ dose = \frac{energy}{mass} \qquad D = \frac{E}{m}$$

The gray, Gy, is the unit of absorbed dose.

One gray is one joule per kilogram: $1\,\text{Gy} = 1\,\text{J}\,\text{kg}^{-1}$

Example 1

A man has his whole body irradiated with alpha radiation until he has received 1 J of energy. Calculate his absorbed dose. Assume his mass is 80 kg.

$$D = \frac{E}{m} = \frac{1}{80} = 0{\cdot}0125\,\text{Gy} = 12{\cdot}5\,\text{mGy}$$

For comparison, a chest X-ray might deliver 0·3 mGy.

TOP TIP

Different materials will absorb different amounts of energy from radiation.

Radiation weighting factor

The absorbed dose measures the energy deposited in the tissue but does not take into account the effects of different types of radiation on the human body. The same dose of a different radiation can have a different biological effect.

A radiation weighting factor (w_r) is given to each type of radiation (see Specified Data page11) as a measure of its biological effect. The higher this value, the higher the biological effect.

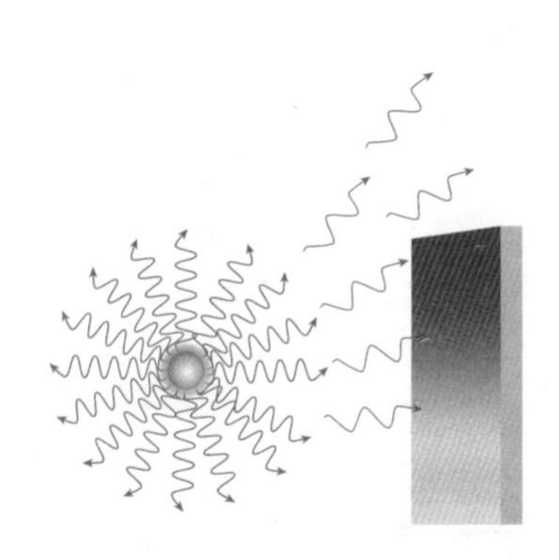

Equivalent dose

Equivalent dose combines the absorbed dose with the type of radiation. The equivalent dose (H) is the product of absorbed dose and radiation weighting factor.

Equivalent dose is measured in sieverts (Sv). $H = Dw_r$

The same equivalent dose always gives the same biological effect. For example, we now know that 1 mSv of γ radiation will do the same damage as 1 mSv of α radiation.

Example 2

The man in example 1 received his absorbed dose from alpha radiation, so his equivalent dose would be: $H = Dw_r = 0{\cdot}0125 \times 20 = 0{\cdot}25\ \text{Sv}$

Example 3

A nuclear industry operator receives an absorbed dose of 300μGy from slow neutrons and an absorbed dose of 4 mGy from gamma radiation.

Calculate the total equivalent dose received.

We need to calculate the equivalent dose from each radiation separately before totalling:

Equivalent dose from slow neutrons

$$H_n = Dw_r = 300 \times 10^{-6} \times 3 = 900 \times 10^{-6} = 0{\cdot}9\ \text{mSv}$$

Equivalent dose from gamma radiation

$$H_g = Dw_r = 4 \times 10^{-3} \times 1 = 4 \times 10^{-3} = 4\ \text{mSv}$$

Total equivalent dose

$$H = H_n + H_g = 0{\cdot}9 + 4 = 4{\cdot}9\ \text{mSv}$$

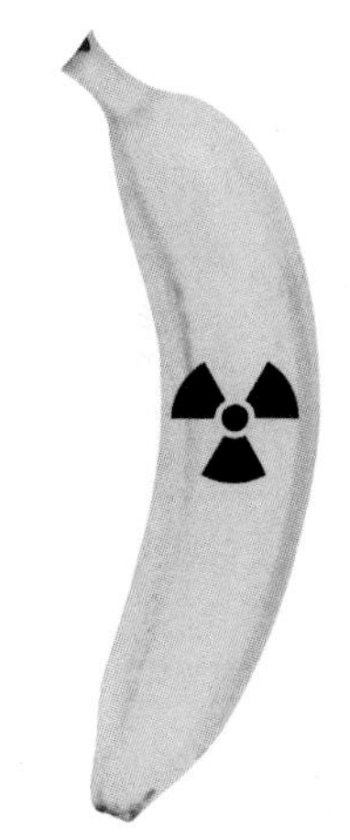

One not to worry over: due to Potassium-40, a banana gives you an equivalent dose of 0·1 μSv.

A further factor that affects the biological risk is the tissue type that the radiation is absorbed by. Doctors will study the tissue weighting factor if only a part of the body receives radiation.

Equivalent dose rate

When equivalent dose is divided by the time for the exposure we get the equivalent dose rate: $\dot{H} = \frac{H}{t}$

0·03μSv is received in 6 hours. $\dot{H} = \frac{H}{t} = \frac{0{\cdot}03 \times 10^{-6}}{6} = 5 \times 10^{-9}\ Svh^{-1}$

Quick Test 50

1. State what is meant by absorbed dose.
2. What is a radiation weighting factor?
3. What is equivalent dose?
4. Calculate the dose when 0·2 J is absorbed by 5 kg of tissue.
5. Alpha particles give a hospital worker an absorbed dose of 5 μGy. Calculate the equivalent dose.

Half-life

Physics to learn	Difference between activity and count rate, half-life.
Revision guide	You can measure count-rate and measure half-life from experimental data. Create and study half-life graphs.

Detecting ionising radiations

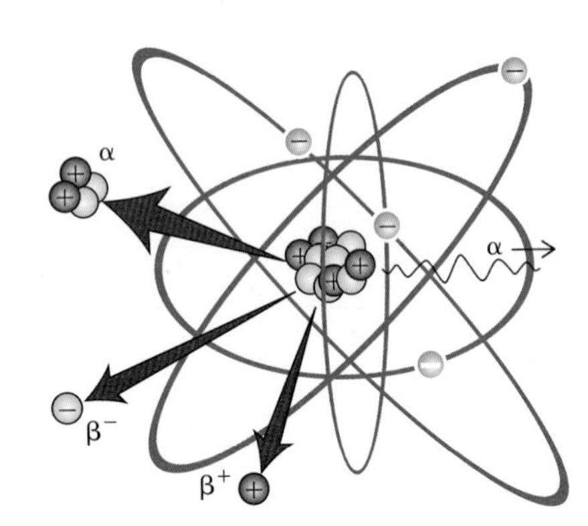

alpha and beta particles and gamma photon

Nuclear radiations can be detected with a Geiger counter (above left). When in use we are not measuring the total activity of the source but are taking a sample measurement of the radioactivity that enters the Geiger-Müller tube of the counter.

This is recorded as a count rate and is measured in counts per minute or counts per second or becquerels (Bq). 1 cs^{-1} = 1 Bq.

Activity and half-life

When a radioactive source emits ionising radiations [alpha (α), beta (β) and gamma (γ)] its nuclei are disintegrating or decaying and the source is changing into a different substance. The activity depends on the number of nuclei in the source: more nuclei = > more activity. The activity is the number of disintegrations occurring per second. The activity of a radioactive source decreases with time.

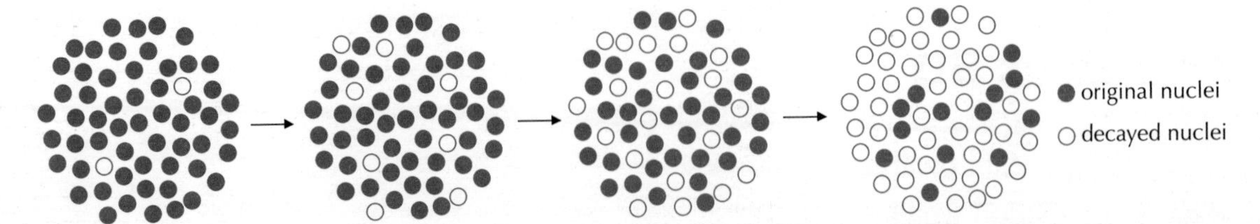

As the nuclei disintegrate, there are fewer nuclei left to emit radiations.

The time taken for half the nuclei in a radioactive substance to decay does not change. This time is the half-life. The half-life of a radioactive substance is the time taken for the activity to drop to half its original value.

TOP TIP

The half-life time can be taken from any two points on a graph where the count rate halves.

Radioactivity is a random process. The decay of an individual nucleus cannot be predicted. The decay in activity can only be predicted because there are large numbers of nuclei in any sample. Different substances have different half-life times.

Nuclear medicine	Sodium 24	15 hours
	Iodine 131	8 days
	Cobalt 60	5·3 years
Carbon dating	Carbon 14	5760 years
Ageing rocks	Uranium 238	4500 million years

Measuring half-life

To measure the half-life of a source you will need:

- a detector of radiation, e.g. a Geiger-Müller tube and counter
- a stopwatch

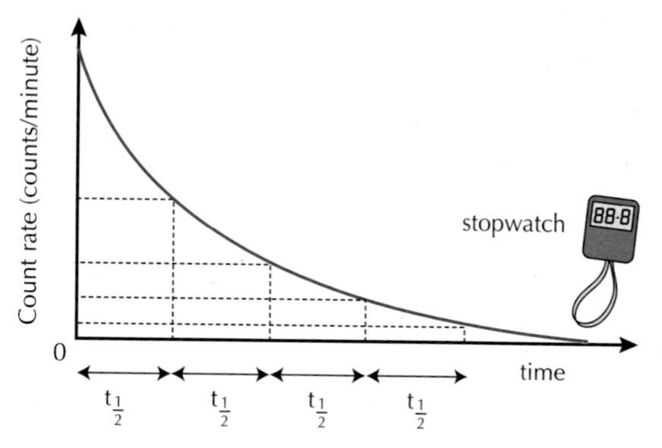

1. First record the background count several times. Calculate the average and subtract this value from all readings.
2. Place the source a fixed distance in front of the Geiger counter and record the count rate at regular time intervals.
3. Plot the count rate (corrected for background count) against the time taken on a graph.
4. Measure the time taken from any initial value of count rate on the graph to half this value.
5. The time for the count rate to keep halving stays the same. This is the half-life time.

Calculations

The fractional activity is useful in understanding how the activity changes with time. Remember, a half-life has not taken place until the activity has halved: start with zero half-lives as shown in this table.

Number of half-lives	Activity (fraction)
Start = 0	1
1	1/2
2	1/4
3	1/8
4	1/16

TOP TIP

Most half-life questions are one of the three types of examples given on these pages. So study them well!

Example 1

Calculate the half-life of the source whose count rate has been recorded in the following table. The background count was recorded at 30 counts per minute.

Time (minutes)	0	30	60	90	120	150
Counts per minute	834	580	429	288	232	175
Corrected count rate (c/m)	*804*	*550*	*399*	*258*	*202*	*145*

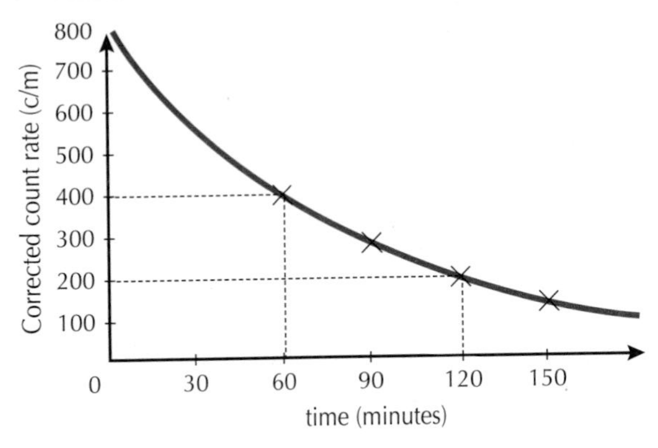

From the graph, the half-life of this source is 60 minutes.

Example 2

A source of activity 12 000 kBq has a half-life of four weeks. It is locked in a cupboard for 16 weeks. What is its activity after this time?

Activity	Number of half-lives
12 000	0
6000	1
3000	2
1500	3
750	4

Number of half-lives $= \frac{16}{4} = 4$

12 000 –> 6000 –> 3000 –> 1500 –> 750

1 2 3 4

The activity after 16 weeks is 750 kBq.

Example 3

The activity of a radioactive source is 1600 MBq. 120 minutes later its activity is only 100 MBq. What is the half-life of the source?

Activity	Number of half-lives
1600	0
800	1
400	2
200	3
100	4

16 000 –> 800 –> 400 –> 200 –> 100

1 2 3 4

There are four half-lives.

$\frac{120}{4} = 30$ minutes

TOP TIP

How many half-lives? Count the changes or arrows.

Quick Test 51

1. State the meaning of the term half-life.
2. What is needed to measure half-life?
3. Activity drops from 1200 Bq to 300 Bq in 30 minutes. What is the half-life?
4. A source with activity 2400 Bq has a half-life of 5 s. Calculate the activity after 25 s.

Nuclear fission

Physics to learn	Nuclear fission.
Revision guide	Describe fission, chain reaction and nuclear power for energy.

Nuclear fission

Fission is the splitting of a large nucleus. In fission, a nucleus with a large mass number splits into two nuclei of smaller mass numbers, usually with the release of neutrons. Energy is released.

slow neutron

$^{235}_{92}U$

$^{141}_{56}Ba$

$^{92}_{36}Kr$

Fission may be induced by neutron bombardment. An incident neutron can stimulate the fission of a nucleus with a large mass number. In the following reaction, the U^{235} momentarily becomes U^{236}, but this is unstable and immediately undergoes fission.

$$^{1}_{0}n + ^{235}_{92}U \rightarrow ^{141}_{56}Ba + ^{92}_{36}Kr + 3^{1}_{0}n + \text{energy}$$

Induced fission is used in the reactors in nuclear power stations.

Chain reaction

The neutrons released during fission can set off a chain reaction. Each neutron released can be made to cause a further fission reaction. The reaction needs to be controlled.

Control rods

Electrical energy supply is maintained if one neutron from each fission reaction causes a further fission. This happens in a controlled chain reaction. Control rods made of boron can move in between the fuel rods to absorb neutrons. Rods can be partially raised to meet demand or fully put in place in an emergency.

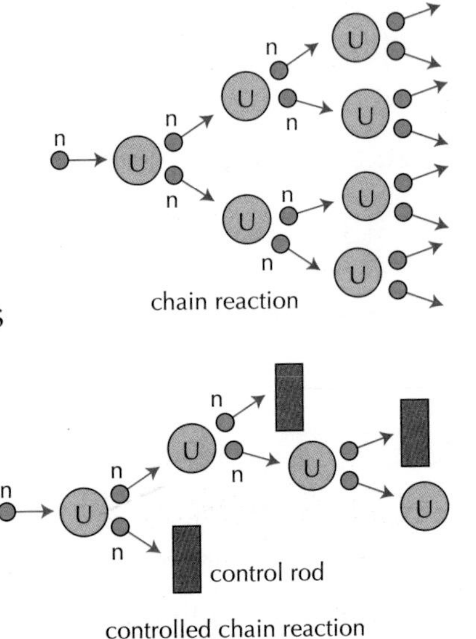

chain reaction

controlled chain reaction

Nuclear power by fission

Today's nuclear power stations all use nuclear fission.

The main parts of a nuclear power station are the nuclear reactor followed by a turbine and generator. The nuclear reactor uses fission of a fuel such as uranium to produce large amounts of heat energy. The heat is used to make steam to drive the turbine and generator. Electrical energy is produced.

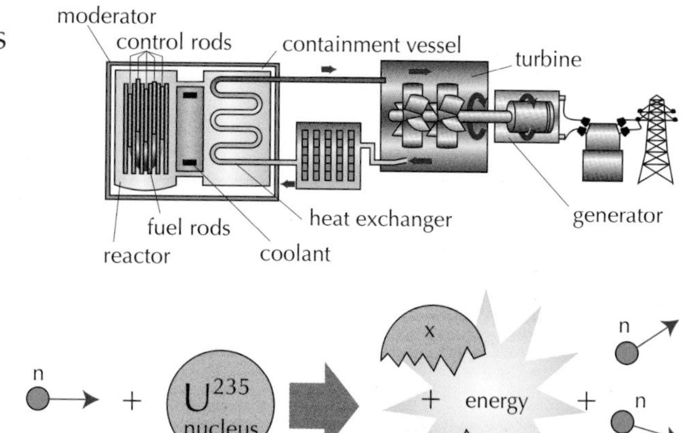

In the reactor, nuclear fuel (uranium) is bombarded with neutrons. Splitting the nucleus is the process of nuclear fission. The uranium 235 nucleus absorbs a neutron, then the nucleus splits into two fission fragments plus two or three neutrons and a large amount of energy. This energy drives the power station.

In nuclear fission, the three neutrons released travel too fast to cause further fission, so in a nuclear reactor they are slowed first by travelling through a moderator. Control rods are also used to absorb neutrons to control the chain reaction.

TOP TIP

You can use the internet to research further the fission process in nuclear power stations.

Quick Test 52

1. What is nuclear fission?
2. What is a chain reaction?
3. Name two parts that are common in most power stations.

Nuclear fusion

Physics to learn	Nuclear fusion.
Revision guide	Describe fusion, plasma containment and the search for fusion reactors.

Nuclear fusion

Fusion is the joining of nuclei. In fusion, two nuclei combine to form a nucleus of larger mass number. The nuclei that fuse together are usually very small. A large amount of energy is released, and no radioactive waste is produced in the reaction.

Fusing light nuclei together releases huge amounts of energy. On earth we have plenty deuterium in sea water and tritium can be made from lithium.

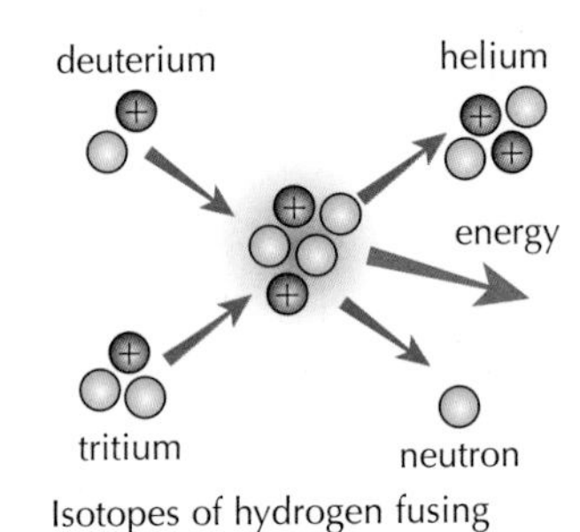

Isotopes of hydrogen fusing

$${}^{2}_{1}\text{H} + {}^{3}_{1}\text{H} \rightarrow {}^{4}_{2}\text{He} + {}^{1}_{0}\text{n} + \text{energy}$$

Plasma containment

To bring nuclei together for fusion reactions is hard, due to the repulsive nature of the positive charge in the nuclei. Hugh amounts of energy have to be found which will bring these nuclei together yet will be less than the energy that we expect to be released!

One thought is that high temperatures are required to simulate fusion which takes place in our sun. At high temperature molecules lose their electrons and become positively charged ions, a plasma. Hydrogen will exist as a plasma, not a gas.

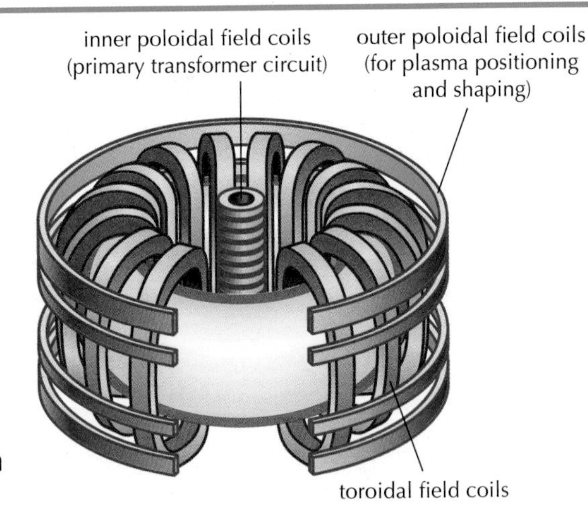

The plasma must be contained so that fusion will occur, energy can be released and more fuel injected.

One design is to contain the plasma in a circular magnetic bottle. Ordinary materials cannot contain the plasma due to the high temperatures required for the fusion.

The plasma must be:

- heated by inducing high electric currents within it
- contained to keep it away from the walls which would melt
- confine plasma density for high energy output

The TOKAMAK is a Russian design that was invented in the 1960s and is a toroidal magnetic chamber. The cost of building the first power station even for experimental use was so large that it is now still under construction and involves 36 countries in its finance.

TOP TIP

Learn both nuclear fission and fusion.

Research

Fusion powers the Sun and stars as hydrogen atoms fuse together to form helium, and matter is converted into energy. The research into creating fusion reactors for power involves having international experiments which are the biggest taking place on earth.

We have examined magnetic confinement fusion as one of the leading methods being studied for nuclear fusion, in the TOKAMAK or torus shaped magnetic chamber.

ITER (the International Thermonuclear Experimental Reactor) is being built in southern France. ITER is the world's largest fusion experiment.

There is also inertial confinement fusion, which usually involves laser beams being focussed onto the surface of the nuclear fuels, such as deuterium and tritium.

Many other methods of undertaking the conditions for nuclear fusion are under research and as this is a current development in physics you should be involved in carrying out some research into the development of nuclear power for the generation of energy.

Quick Test 53

1. What is the biggest challenge for nuclear fusion?
2. What is nuclear fusion?

Units, prefixes and scientific notation

Physics to learn	National 5 scientific requirements.
Revision guide	Identify and use appropriately.

Units

The basic SI units that you will have used in this course are the *metre* for measurement of length, the *kilogram* for mass, the *second* for time, the *ampere* for electric current, the *kelvin* for temperature.

Quantity	Name	Symbol
length	metre	m
mass	kilogram	kg
time	second	s
current	ampere	A
temperature	kelvin	K

TOP TIP

Note, the symbol is capitalised when a name is used.

In addition we use what are known as derived units. These include:

Quantity	Name	Symbol
area	square metre	m^2
volume	cubic metre	m^3
velocity	metre/second	$m \cdot s^{-1}$
acceleration	metre/second squared	$m \cdot s^{-2}$
frequency	hertz	Hz
force	newton	N
energy, work	joule	J
power	watt	W
charge	coulomb	C
voltage, p.d.	volt	V
resistance	ohm	Ω
activity	becquerel	Bq
absorbed dose	gray	Gy
equivalent dose	sievert	Sv

TOP TIP

Remember to include the correct unit in your final answer.

prefixes

nano	n	0·000 000 001	10^{-9}
micro	μ	0·000 001	10^{-6}
milli	m	0·001	10^{-3}
kilo	k	1000	10^{3}
mega	M	1000 000	10^{6}
giga	G	1000 000 000	10^{9}

TOP TIP

Remember to check the pre-fixes in the exam questions.

Scientific notation

Scientific notation helps us deal with extremely large or small numbers often found in Physics:

Examples

Multiplication: $(3 \times 10^4) \times (2 \times 10^3) = (3 \times 2) \times (10^{4+3}) = 6 \times 10^7$

Division: $(3 \times 10^4) / (2 \times 10^3) = (3/2) \times (10^{4-3}) = 1{\cdot}5 \times 10^1 = 15$

Significant figures

If a laboratory trolley travels across the bench we can attempt to measure its average speed:

Suppose we put two marks down, say 1m apart. We then attempt to time the trolley as it travels between these two marks with a stop-clock which only reads seconds. Our calculator will do the following calculation:

$$v = \frac{d}{t} = \frac{1}{3} = 0{\cdot}333333\text{ms}^{-1}.$$

However suppose we could more accurately measure the marks as 102 cm and the time as 2·99s then the calculation gives 0·341137 ms^{-1}.

The final answer can have no more significant figures than the value with least number of significant figures used in the calculation.

Our two answers should say $v = 0{\cdot}3\text{ms}^{-1}$ and $v = 0{\cdot}341\text{ ms}^{-1}$.

TOP TIP

Only give an appropriate number of significant figures in our final answer.

Quick Test 54

1. In prefixes, state the difference between m and M.

Exam questions guide

Physics to learn	Exam questions.
Revision guide	Objective and extended response style.

Objective questions

The first section of your exam paper will have objective or multiple choice questions. You must only select one answer for each question.

Here are some questions from different areas of physics to practice:

1. Which of the following contains a scalar followed by a vector quantity?
   A) speed and time
   B) mass and energy
   C) speed and force
   D) velocity and acceleration
   E) acceleration and velocity
2. Studying the spectrum of light from the stars is known as
   A) Dynamics
   B) Electricity
   C) Physics
   D) Spectroscopy
   E) Capacity
3. A 6V battery drives a current of 2·5A in a circuit. What is the circuit resistance value?
   A) 15 Ω
   B) 8·5 Ω
   C) 3·5 Ω
   D) 2·5 Ω
   E) 2·4 Ω
4. Water to make tea has to rise from 20°C to 100°C. What is the temperature rise in kelvin?
   A) 20 K
   B) 60 K
   C) 100 K
   D) 80 °C
   E) 80 K
5. Waves travel across a pond. Their wavelength is measured as 0·5m and their frequency is 4Hz. What is their speed?
   A) 2 $m.s^{-1}$
   B) 2 Hz
   C) 4·5 $m.s^{-1}$
   D) 8 $m.s^{-1}$
   E) 0·125 $m.s^{-1}$
6. A biological material receives an absorbed dose of 200 mGy from slow neutrons. What is the value of the equivalent dose?
   A) 200 mSv
   B) 600 mSv
   C) 600 mGy
   D) 67 mSv
   E) 200 nGy

TOP TIP

Remember to only select one answer per question.

Extended response questions

The second section of your exam paper will require extended responses.

You will write your responses throughout the paper.

1. 30 000 decays are emitted over a 15 s time. What is the activity of this nuclear source?

   (Note that this is a standard type question. You should identify the equation to use, give the substitution values and provide the answer with the correct unit).

2. A 2kW kettle heats 0·18 kg of water from 20 °C to 100 °C. What is the minimum time that this will take?

3. A vehicle accelerates at 3 $m.s^{-2}$ from 10 $m.s^{-1}$ for a period of 5s.

   A) Calculate the final speed the vehicle should reach.

   B) Explain whether the vehicle's final speed will be greater or less than this value.

4. Calculate the power being developed in this lamp.

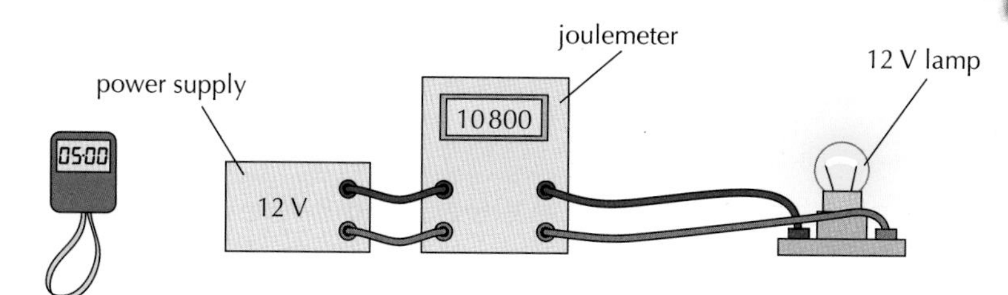

5. An object undertakes the motion shown on the graph. Show that the displacement of this object is 320 m.

   (Note that here it is essential that you show the full steps required to obtain the answer, it is only the steps that earn the marks).

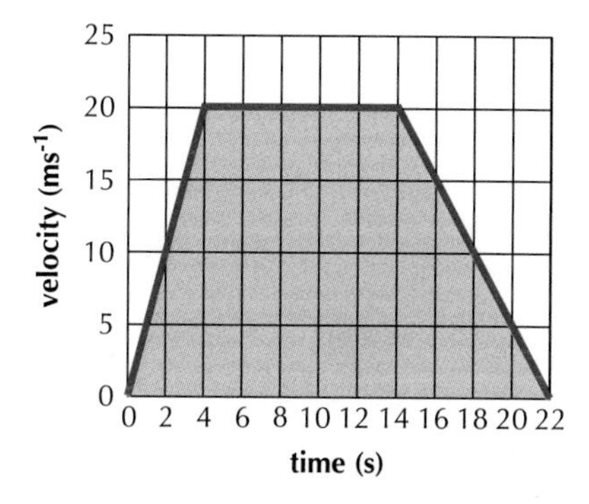

TOP TIP

You can do the questions you find easiest first.

TOP TIP

Practice past papers from the SQA as you approach your exam.

In the exam you will find a variety of questions. You will find questions based on course knowledge, experimental knowledge and skills, and you will have to apply your learning to other situations. It is important that you study the diagrams and illustrations as well as the text as you revise with this success guide.

National 5 PHYSICS

For SQA 2019 and beyond

Practice Papers

Michael Murray

Introduction

The two papers included in this section are designed to provide practice in the National 5 Physics Course assessment question paper (the examination), which is worth 80% of the final grade for this course.

Together, the two papers give overall and comprehensive coverage of the assessment of **Skills, Knowledge and Understanding** needed to pass National 5 Physics.

We recommend that candidates download a copy of the Physics Course Specification from the SQA website at www.sqa.org.uk. Print pages 5–13, which summarise the knowledge and skills that will be tested.

Design of the papers

Each paper has been carefully assembled to be very similar to a typical National 5 question paper. Each paper has 135 marks and is divided into two sections.

- **Section 1** – Objective Test, which contains 25 multiple choice items worth 1 mark each and totalling 25 marks altogether.
- **Section 2** – Paper 2, which contains restricted and extended response questions totalling 110 marks altogether.

A data sheet containing relevant data is provided with each paper along with a list of all formulae. Candidates should familiarise themselves with the data and formulae that will be made available to them during the exam. Data and relationship sheets can be found on pages 138 and 139 respectively in this book and should be consulted when required.

Most questions in each paper are set at the standard of Grade C but there are also more difficult questions set at the standard for Grade A. We have attempted to construct each paper to represent the typical range of demand in a National 5 Physics paper.

Expected answers

The expected answers online at www.leckieandleckie.co.uk give National Standard answers but, occasionally, there may be other acceptable answers. For example, when giving a numerical answer there will be a range of significant figures that will be acceptable. As a rough guide, try and give answers to a similar number of significant figures as the data given in the question.

There are Top Tips provided alongside each answer. These include hints on the Physics itself as well as some memory ideas, a focus on traditionally difficult areas, advice on the wording of answers and notes of commonly made errors.

Grading

The two papers are designed to be equally demanding and to reflect the National Standard of a typical SQA paper. Each paper has 135 marks – if you score 68 marks that's a C pass. You will need about 81 marks for a B pass and about 95 marks for an A. **These figures are a rough guide only.**

Timing

If you are attempting a full paper, limit yourself to **2 hours 30 minutes** to complete it. We recommend around 30 minutes for the Objective Test and the remainder of the time for Section 2.

For extended response questions give yourself about a minute per mark, for example, a 10-mark question should take no longer than around 10 minutes.

Good luck!

Data Sheet

Speed of light in materials

Material	Speed in m s^{-1}
Air	$3{\cdot}0 \times 10^8$
Carbon dioxide	$3{\cdot}0 \times 10^8$
Diamond	$1{\cdot}2 \times 10^8$
Glass	$2{\cdot}0 \times 10^8$
Glycerol	$2{\cdot}1 \times 10^8$
Water	$2{\cdot}3 \times 10^8$

Gravitational field strengths

	Gravitational field strength on the surface in N kg^{-1}
Earth	9·8
Jupiter	23.0
Mars	3·7
Mercury	3·7
Moon	1·6
Neptune	11.0
Saturn	9·0
Sun	270.0
Uranus	8·7
Venus	8·9

Specific latent heat of fusion of materials

Material	Specific latent heat of fusion in J kg^{-1}
Alcohol	$0{\cdot}99 \times 10^5$
Aluminium	$3{\cdot}95 \times 10^5$
Carbon Dioxide	$1{\cdot}80 \times 10^5$
Copper	$2{\cdot}05 \times 10^5$
Iron	$2{\cdot}67 \times 10^5$
Lead	$0{\cdot}25 \times 10^5$
Water	$3{\cdot}34 \times 10^5$

Specific latent heat of vaporisation of materials

Material	Specific latent heat of vaporisation in J kg^{-1}
Alcohol	$11{\cdot}2 \times 10^5$
Carbon Dioxide	$3{\cdot}77 \times 10^5$
Glycerol	$8{\cdot}30 \times 10^5$
Turpentine	$2{\cdot}90 \times 10^5$
Water	$22{\cdot}6 \times 10^5$

Speed of sound in materials

Material	Speed in m s^{-1}
Aluminium	5200
Air	340
Bone	4100
Carbon dioxide	270
Glycerol	1900
Muscle	1600
Steel	5200
Tissue	1500
Water	1500

Specific heat capacity of materials

Material	Specific heat capacity in J kg^{-1} °C^{-1}
Alcohol	2350
Aluminium	902
Copper	386
Glass	500
Ice	2100
Iron	480
Lead	128
Oil	2130
Water	4180

Melting and boiling points of materials

Material	Melting point in °C	Boiling point in °C
Alcohol	−98	65
Aluminium	660	2470
Copper	1077	2567
Glycerol	18	290
Lead	328	1737
Iron	1537	2737

Radiation weighting factors

Type of radiation	Radiation weighting factor
alpha	20
beta	1
fast neutrons	10
gamma	1
slow neutrons	3

Relationship Sheet

$E_p = mgh$

$E_k = \frac{1}{2}mv^2$

$Q = It$

$V = IR$

$R_T = R_1 + R_2 + ...$

$\frac{1}{R_T} = \frac{1}{R_1} + \frac{1}{R_2} + ...$

$V_2 = \left(\frac{R_2}{R_1 + R_2}\right)V_s$

$\frac{V_1}{V_2} = \frac{R_1}{R_2}$

$P = \frac{E}{t}$

$P = IV$

$P = I^2R$

$P = \frac{V^2}{R}$

$E_h = cm\Delta T$

$p = \frac{F}{A}$

$\frac{pV}{T} = \text{constant}$

$p_1V_1 = p_2V_2$

$\frac{p_1}{T_1} = \frac{p_2}{T_2}$

$\frac{V_1}{T_1} = \frac{V_2}{T_2}$

$d = vt$

$v = f\lambda$

$T = \frac{1}{f}$

$A = \frac{N}{t}$

$D = \frac{E}{m}$

$H = Dw_R$

$\dot{H} = \frac{H}{t}$

$s = vt$

$d = \bar{v}t$

$s = \bar{v}t$

$a = \frac{v - u}{t}$

$W = mg$

$F = ma$

$E_w = Fd$

$E_h = ml$

Practice paper A

N5 Physics

Practice Papers for SQA Exams

Physics Section 1

Fill in these boxes:

Name of centre

Town

Forename(s)

Surname

Try to answer all of the questions in the time allowed.

Total marks — 135

Section 1 — 25 marks

Section 2 — 110 marks

Read all questions carefully before attempting.

You have 2 hours 30 minutes to complete this paper.

Write your answers in the spaces provided, including all of your working.

SECTION 1

Objective Test

1. Which of the following contains two vector quantities?

A Distance and speed

B Displacement and speed

C Displacement and velocity

D Weight and mass

E Force and mass

2. A student sets up the apparatus as shown.

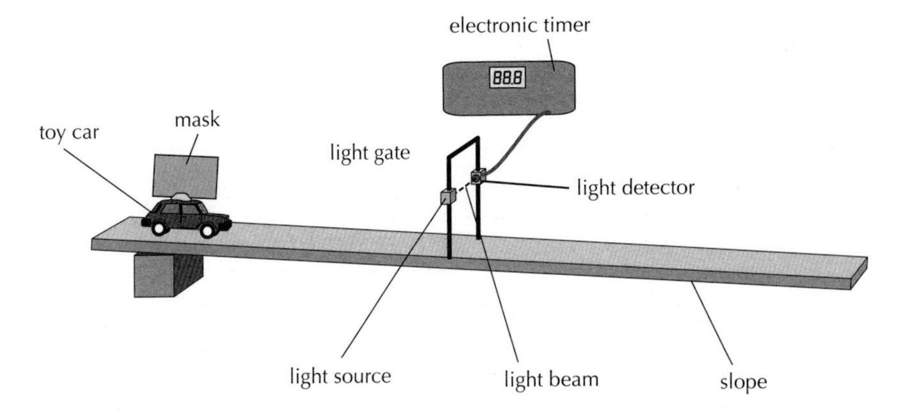

The toy car is released from rest at the top of the slope.

The following measurements are recorded.

Time for mask to pass through light gate = 0·04 s.

Time taken for toy car to travel from top of slope to light gate = 0·25 s.

Length of mask = 50 mm.

The instantaneous speed of the toy car as it passes through the light gate is

A 0·20 m s^{-1}

B 0·80 m s^{-1}

C 1·25 m s^{-1}

D 5·00 m s^{-1}

E 20·0 m s^{-1}.

3. The graph shows how the velocity of an object varies with time

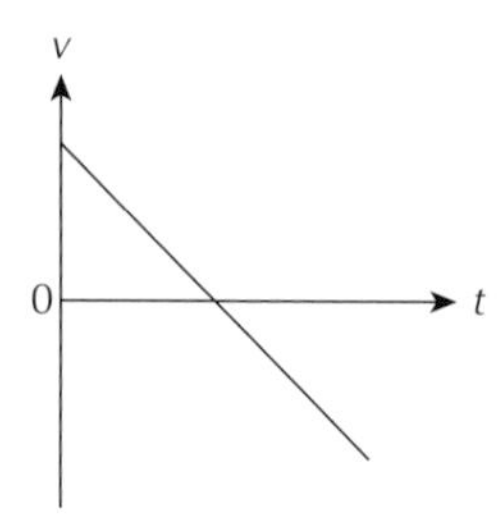

The graph could represent the motion of

A a rocket accelerating upwards

B a ball thrown upwards which then falls back towards the ground

C a person decelerating along a track then accelerating in the same direction

D a car decelerating then accelerating

E a ball falling downwards

4. Three objects, X, Y and Z, travel in a straight line. The table below shows the velocities of the three objects for a time interval of 3 seconds.

Time **(s)**	0	1	2	3
Velocity of X **(m s⁻¹)**	0	1	2	3
Velocity of Y **(m s⁻¹)**	0	1	3	5
Velocity of Z **(m s⁻¹)**	4	6	8	10

Which of the following statements is/are correct?

I X moves with constant acceleration.

II Y moves with constant acceleration.

III Z moves with constant velocity.

A I only

B II only

C III only

D I and III only

E I, II and III

5. An astronaut has a mass of 70 kg. Which row of the table gives the mass and weight of the astronaut on the Moon?

	Mass **(kg)**	***Weight*** **(N)**
A	112	112
B	112	70
C	70	70
D	70	112
E	70	686

6. A person sits on a chair that rests on the Earth. The person exerts a downward force on the chair.

Which of the following completes the 'Newton pair' of forces?

A The force of the person on the Earth.

B The force of the person on the chair.

C The force of the chair on the person.

D The force of the chair on the Earth.

E The force of the Earth on the person.

7. A car of mass 1000 kg is travelling at a speed of 30 m s^{-1}. The brakes are applied and the car decelerates to 10 m s^{-1}.

The change in the car's kinetic energy is

A 10 kJ

B 200 kJ

C 400 kJ

D 445 kJ

E 800 kJ.

8. When a space craft enters the Earth's atmosphere, one effect of friction is to transform

A heat into kinetic energy

B potential energy into heat

C kinetic energy into potential energy

D potential energy into kinetic energy

E kinetic energy into heat

9. It takes light from the Sun 8 minutes to reach the Earth. The distance from the Earth to the Sun in metres is

A $1{\cdot}63 \times 10^5$ m

B $6{\cdot}25 \times 10^5$ m

C $4{\cdot}00 \times 10^7$ m

D $2{\cdot}40 \times 10^9$ m

E $1{\cdot}44 \times 10^{11}$ m

10. An electric heater has a rating of 1·25 kW, 50 Ω.

The charge passing through the element of the heater in 100 s is

A 2·5 C

B 50 C

C 500 C

D 2500 C

E 25 000 C.

11. The voltage of an electricity supply is a measure of the

A power developed in the circuit

B current in the circuit

C resistance of the circuit

D energy given to the charges in the circuit

E speed of the charges in the circuit.

12. Three resistors are connected as shown.

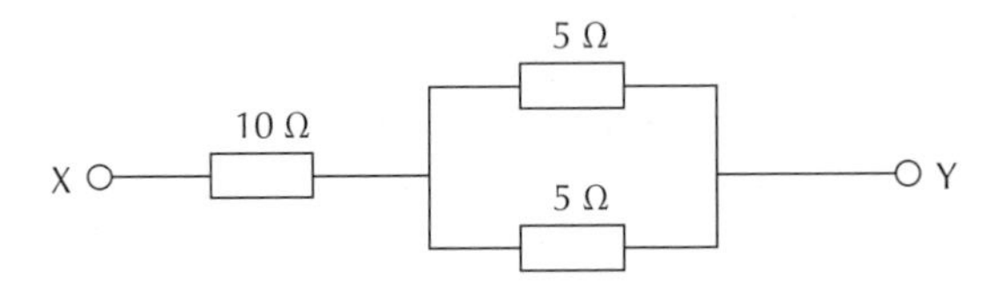

The resistance between X and Y is

A 2·00 Ω

B 10·4 Ω

C 12·5 Ω

D 15·0 Ω

E 20·0 Ω.

13. A 2·2 kW heater is connected to a 230 V mains supply of frequency 50 Hz. The current in the heater is

A 0·10 A

B 9·60 A

C 105 A

D 110 A

E 506 A

14. Two resistors are connected in series with a 54 volt d.c. supply.

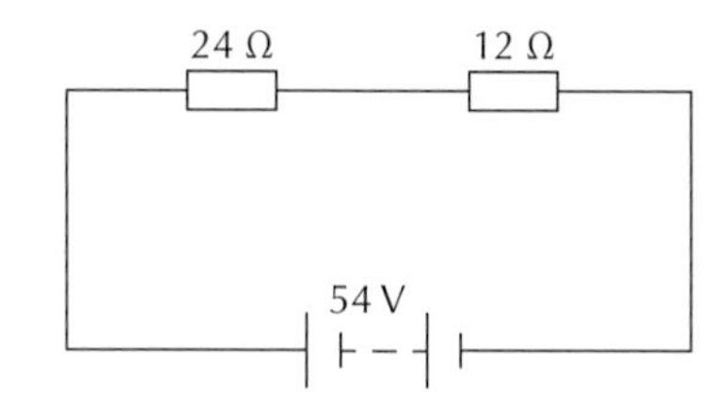

The current in the 24 Ω resistor is 1·5 A.

Which row of the table shows the current in the 12 Ω resistor and the voltage across the 12 Ω resistor?

	Current **(A)**	***Voltage*** **(V)**
A	0·75	9
B	1·5	18
C	1·5	54
D	3·0	36
E	3·0	54

15. A solar furnace holds 150 kg of water.

The water is heated from 25 °C to 85 °C.

The specific heat capacity of water is 4180 J kg^{-1} $°C^{-1}$.

The heat energy gained by the water is

A 9000 J

B 627 000 J

C 15 675 000 J

D 37 620 000 J

E 53 295 000 J.

16. The pressure–volume graph below describes the behaviour of a constant mass of gas when it is heated.

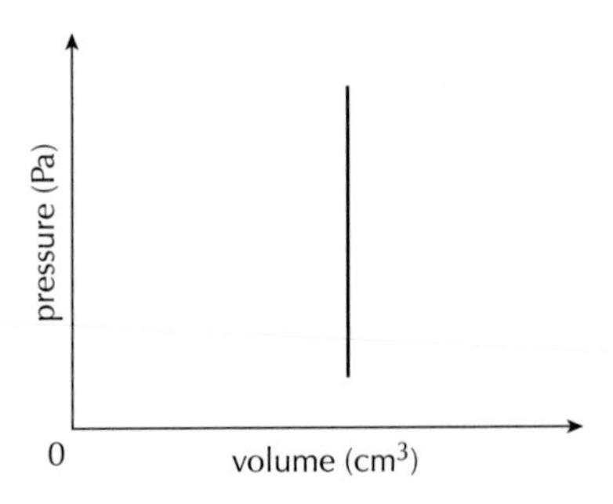

Which of the following shows the corresponding pressure–temperature graph?

A

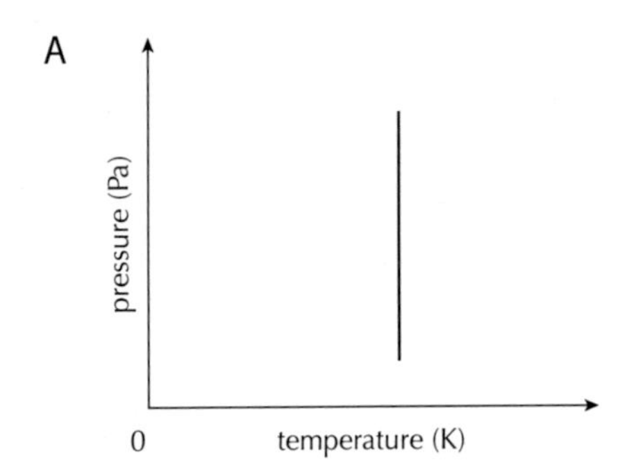

B

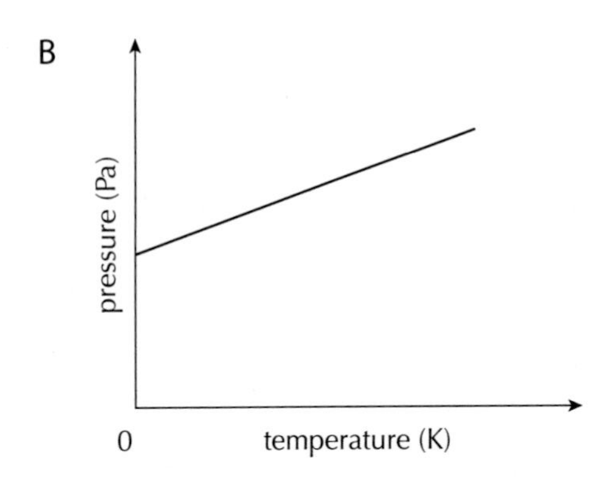

C

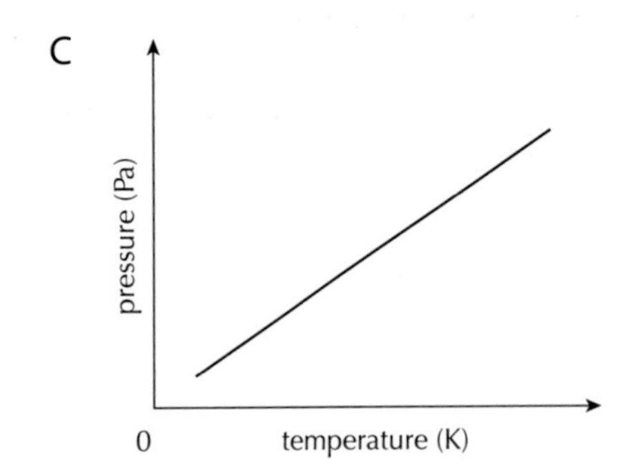

D 

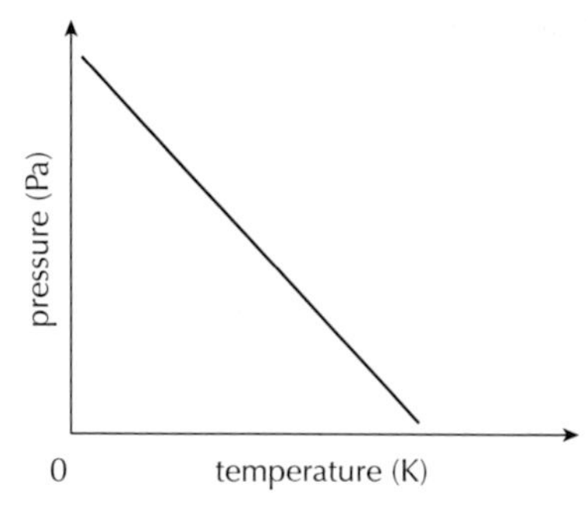

E

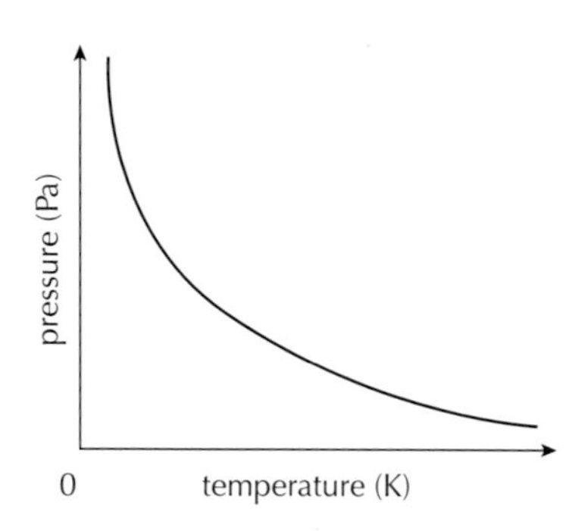

17. Water from a kettle at 90 °C cools to a temperature of 50 °C.

The temperature change on the Kelvin scale is

A 40 K

B 50 K

C 313 K

D 323 K

E 363 K.

18. A radio signal is sent from a transmitter in Scotland to a receiver in Canada.

The distance between transmitter and receiver is $5{\cdot}4 \times 10^3$ km. The signal has a frequency of 100 MHz.

The time taken for the signal to reach the receiver is

A $1{\cdot}8 \times 10^{-8}$ s

B $1{\cdot}8 \times 10^{-5}$ s

C $1{\cdot}6 \times 10^{-2}$ s

D $1{\cdot}8 \times 10^{-2}$ s

E $6{\cdot}0 \times 10^{-2}$ s.

19. In a swimming pool, 15 waves pass a point in 3 seconds. The speed of the waves is 0·6 m s^{-1}. The wavelength of the waves is

A 0·04 m

B 0·12 m

C 0·20 m

D 1·80 m

E 3·00 m.

20. The diagram shows part of the electromagnetic spectrum.

P	Microwaves	Q	Visible light

Identify radiation P and radiation Q.

	P	*Q*
A	radio	ultraviolet
B	radio	infrared
C	infrared	ultraviolet
D	ultraviolet	infrared
E	ultraviolet	radio

21. For a ray of light travelling at an angle from **glass into air**, which of the following statements is/are **true**?

I The speed of light increases.

II The wavelength of light increases.

III The direction of light changes.

A I only

B III only

C I and II only

D I and III only

E I, II and III

22. The diagram shows a ray of light striking a rectangular glass block.

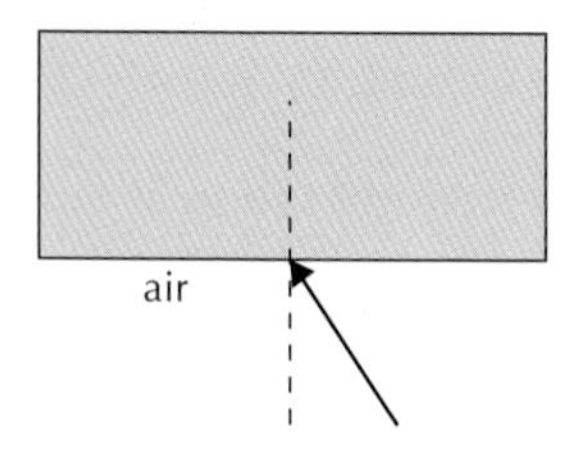

Which diagram shows the path of the ray through the block?

A

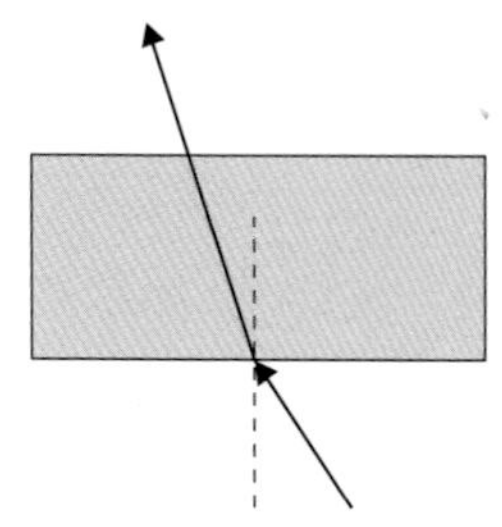

B

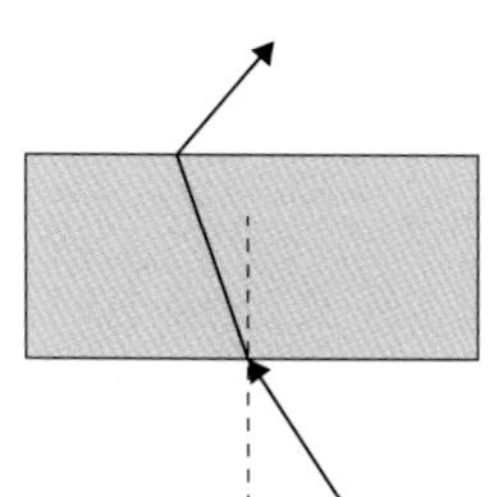

C

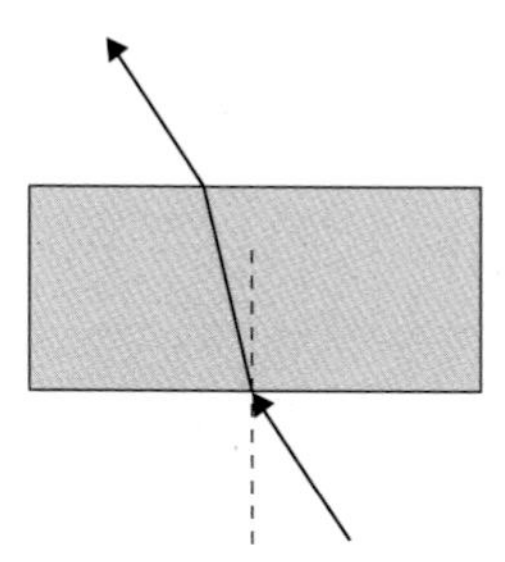

D

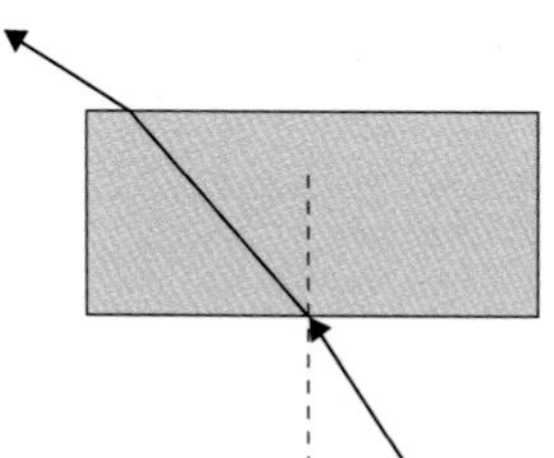

E

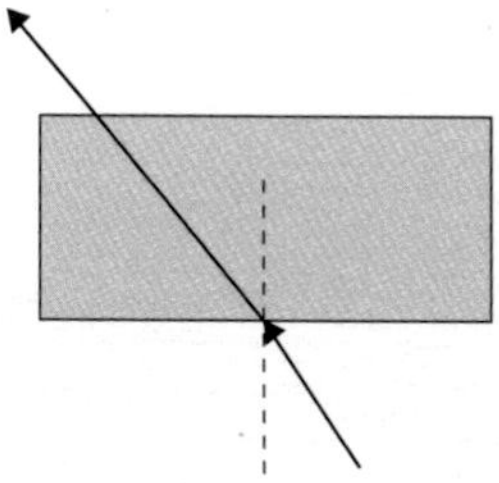

23. A student makes the following three statements.

I Alpha particles have a greater ionisation density than beta particles.

II Alpha particles have a greater ionisation density than gamma rays.

III Gamma rays have a greater ionisation density than alpha particles and beta particles.

Which of the statements is/are **true**?

A I only

B II only

C III only

D I and II only

E I and III only

24. The activity of a radioactive material is 180 Bq. The half-life of the substance is 8 hours.

The time for the activity to fall to 22·5 Bq is

A 4 hours

B 8 hours

C 16 hours

D 20 hours

E 24 hours.

25. Activity can be defined as

A the number of disintegrations per second

B the annual equivalent dose

C the energy per unit mass

D the ionisation density

E the number of sieverts per hour

N5 Physics

Physics Section 2

Fill in these boxes:

Name of centre

Town

Forename(s)

Surname

Try to answer all of the questions in the time allowed.

Total marks — 135

Section 1 — 25 marks

Section 2 — 110 marks

Read all questions carefully before attempting.

You have 2 hours 30 minutes to complete this paper.

Write your answers in the spaces provided, including all of your working.

MARKS
Do not write in this margin

SECTION 2

1. A weather balloon of mass 100 kg rises vertically from the ground.

The graph shows how the vertical velocity of the balloon changes during the first 120 seconds of its upward flight.

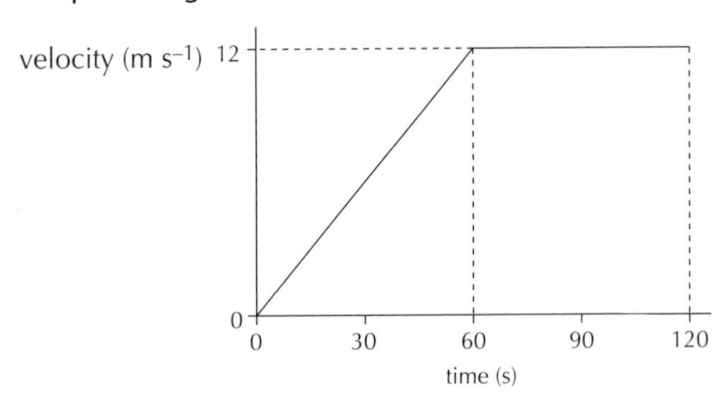

(a) Calculate the initial acceleration of the balloon. **3**

Space for working and answer

(b) Calculate the vertical displacement of the balloon after 120 s. **3**

Space for working and answer

(c) Calculate the kinetic energy of the balloon at 120 s. **3**

Space for working and answer

MARKS
Do not write in this margin

2. An aircraft is transporting skydivers to carry out a parachute jump. The aircraft and passengers have a total mass of 30 000 kg.

The forces exerted on the aircraft are shown on the diagram.

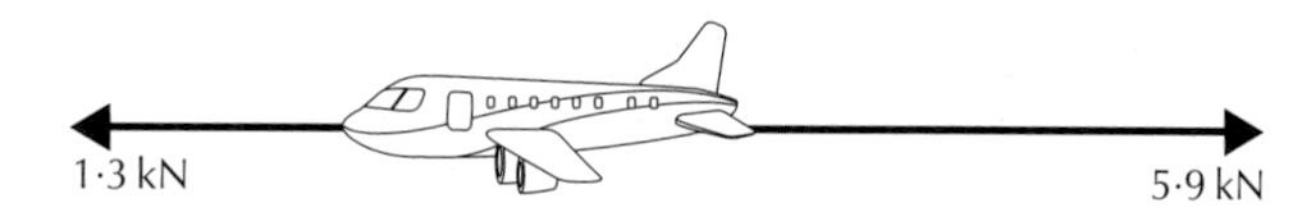

(a) Calculate the acceleration of the aircraft. **3**

Space for working and answer

(b) The aircraft reaches a constant horizontal speed. At a height of 5000 m a skydiver jumps from the aircraft. The skydiver falls for 24 seconds before opening her parachute.

(i) Sketch a diagram showing the forces acting on the skydiver as she falls through the air. You must name the forces and show their direction. **2**

(ii) Calculate the speed of the skydiver the instant before she opens her parachute. Assume the acceleration is constant. **3**

Space for working and answer

MARKS Do not write in this margin

(iii) Explain why the actual speed of the parachutist is much less than the value calculated in part (b) (ii). **2**

(c) A short time after opening her parachute, the skydiver reaches a terminal velocity. Explain what is meant by **terminal velocity**. **1**

(d) Shortly before landing, the skydiver is travelling at $4{\cdot}8\ \text{m s}^{-1}$ horizontally and $2{\cdot}2\ \text{m s}^{-1}$ vertically.

By scale diagram, or otherwise, determine the resultant velocity of the skydiver.

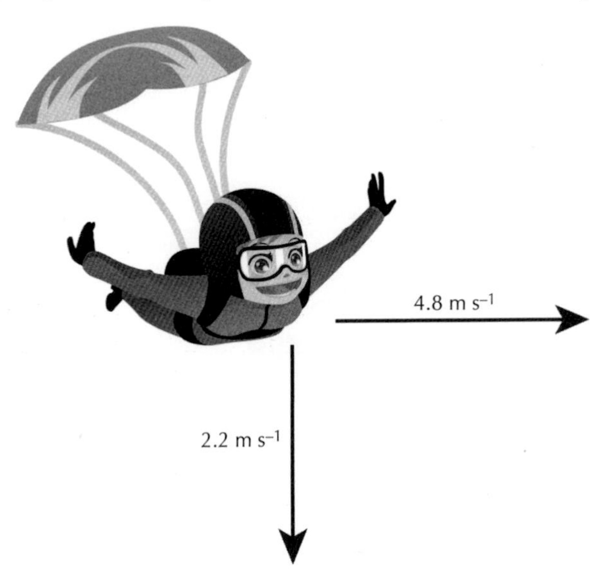

Space for working and answer **4**

MARKS Do not write in this margin

3. A solid aluminium ball of mass 0·75 kg is dropped from the Wallace Monument in Stirling.

(a) (i) The ball is released from a height of 60 m. Calculate the gravitational potential energy lost by the ball.

Space for working and answer **3**

(ii) Assuming that all the gravitational potential energy is converted into heat energy in the ball, calculate the increase in the temperature of the ball on impact with the ground.

Space for working and answer **4**

(b) Is the actual temperature change of the ball less than, the same as of greater than the value calculated in part (a) (ii)?

You must justify your answer. **2**

MARKS
Do not write in this margin

4. *Voyager 1* is a space probe launched by NASA in 1977 to explore the outer planets of the solar system.

NASA communicates with the space probe via radio waves. In 2013, *Voyager 1* became the first man-made object to leave the solar system. The distance from Earth to *Voyager 1* is $1{\cdot}9 \times 10^{10}$ km.

(a) What is a planet? **1**

(b) A radio signal is transmitted from Earth to *Voyager 1*. The signal is processed and 5 minutes later is transmitted back to Earth. Calculate the total time taken for the signal to be sent and received back on Earth.

Space for working and answer **4**

(c) Distances in space are sometimes measured in astronomical units (AU).

1 AU = 149 597 871 km

Calculate the distance from Earth to Voyager 1 in astronomical units.

Space for working and answer **2**

(d) Voyager 1 has travelled further than any other man-made object. Using your knowledge of physics, explain why the Voyager 1 space probe has been able to travel such a large distance from Earth. **2**

5. A science fiction film features an astronaut who becomes detached from a space shuttle in orbit around the Earth. The astronaut spins vertically away from the Earth into the vacuum of space.

Using your knowledge of physics, comment on the motion of the astronaut after becoming detached from the space shuttle. **3**

Space for working and answer

6. A student sets up the following circuit.

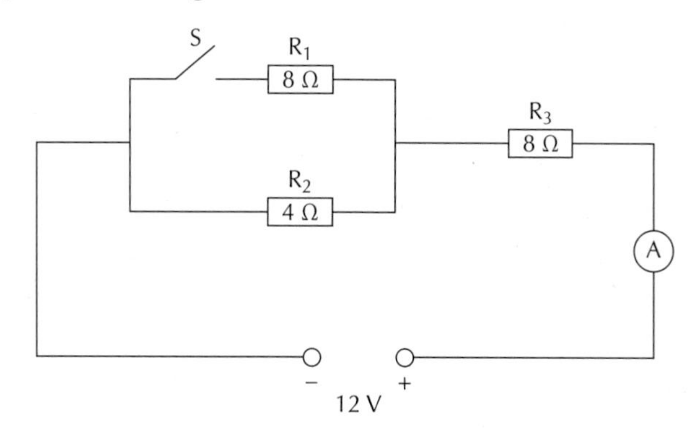

(a) (i) Calculate the current in the circuit when switch S is open. **4**

Space for working and answer

(ii) Calculate the potential difference across R_2 when switch S is closed. **5**

Space for working and answer

(b) The supply voltage is increased to 22 V. R_1 and R_2 are replaced by a 6 V lamp and switch S is removed. The lamp is operating at its stated voltage.

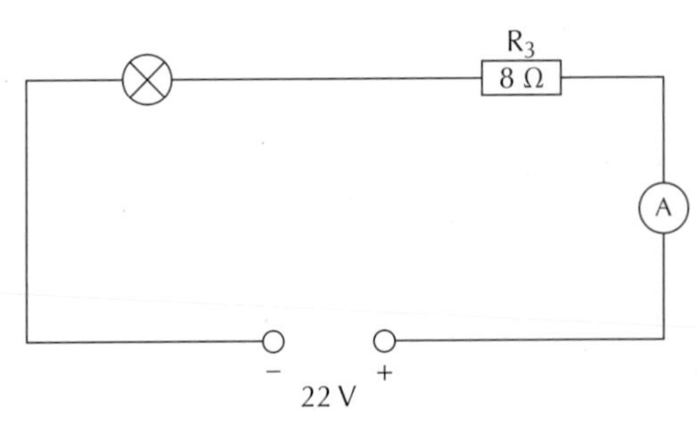

State whether there is a greater power dissipated in R_3 or in the lamp.

Justify your answer by calculation.

Space for working and answer

5

MARKS
Do not write in this margin

7. A mass of 350 g of a substance is heated with a 35 W heater. A temperature probe is inserted into the substance.

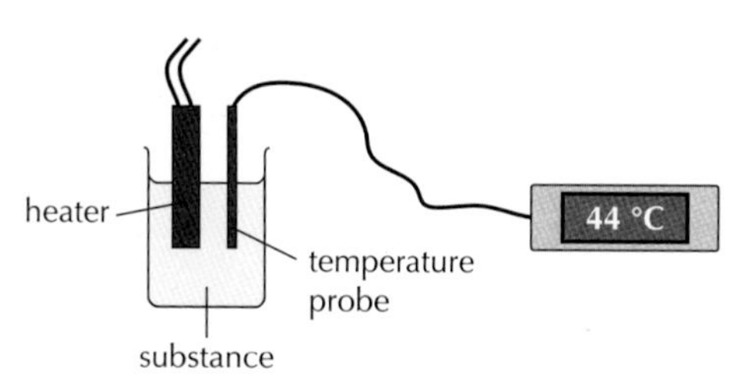

The substance is initially a solid at room temperature. The graph below shows the temperature of the substance from the moment the heater is switched on.

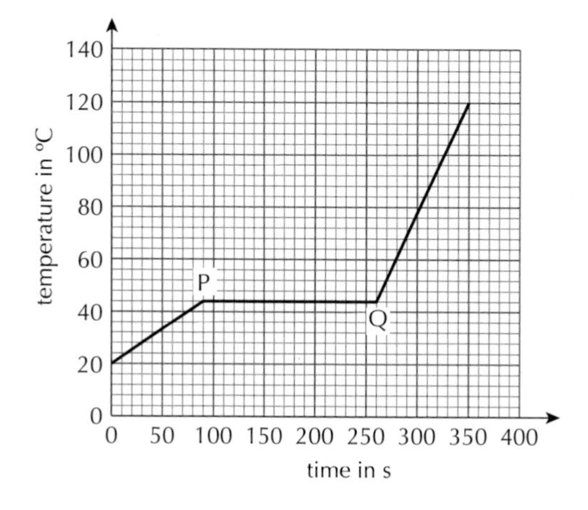

(a) State the value of room temperature **1**

(b) (i) Explain why the temperature of the substance remains constant between P and Q? **1**

(ii) Calculate the energy transferred by the heater during the time interval PQ. **4**

Space for working and answer

(iii) Calculate the specific latent heat of fusion of the substance. **3**

Space for working and answer

MARKS
Do not write in this margin

8. A fire extinguisher of compressed carbon dioxide gas is inside a van.

(a) The carbon dioxide inside the fire extinguisher is at a pressure of $2{\cdot}36 \times 10^6$ Pa and a temperature of 6·0 °C. The cylinder is now moved to a school laboratory where the temperature is 21·0 °C.

(i) Calculate the pressure of the carbon dioxide in the fire extinguisher when its temperature is 21·0 °C. **4**

Space for working and answer

(ii) Use the kinetic model to explain the change in pressure as the temperature increases. **2**

(b) The manufacturer warns against storing the fire extinguisher at temperatures above 48·0 °C.

Explain why a temperature limit must be set. **2**

MARKS
Do not write in this margin

9. Using your knowledge of physics, estimate the pressure you exert on the ground when standing on both feet. **3**

MARKS
Do not write in this margin

10. Manufacturers of solar cells need to know how efficient they are.

Efficiency is defined as the rate at which energy from the Sun is converted to useful output energy.

The efficiency of a solar cell can be determined using:

$$\textit{Percentage efficiency} = \frac{V_{oc} I_{sc}}{P_{in}} \times 100\%$$

where V_{oc} is the open circuit voltage, I_{sc} is the short circuit current and P_{in} is the input power.

The table below gives information on three different solar cells.

	V_{oc} (V)	I_{sc} (A)	P_{in} (W)
Solar cell A	0·7	3·8	10
Solar cell B	0·6	3·5	10
Solar cell C	0·98	2·3	15

(a) Show, by calculation, which solar cell has the highest percentage efficiency. **6**

Space for working and answer

(b) Solar cells use the infrared radiation from sunlight to generate power.

The wavelength of the radiation is 545 nm. Calculate the frequency. **3**

Space for working and answer

MARKS
Do not write in this margin

11. An oven is used to cook popcorn. The oven produces microwaves.

(a) What is transferred by waves? **1**

(b) (i) Are microwaves transverse or longitudinal waves? **1**

(ii) Describe the difference between transverse and longitudinal waves. **2**

(c) Microwaves can also be used to transmit mobile telephone signals.

Complete the diagram below to show the pattern of the microwaves to the right of the hill. **2**

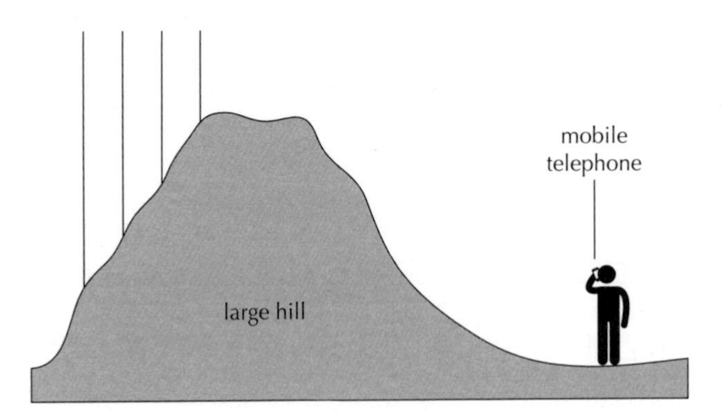

MARKS Do not write in this margin

12. A school technician sets up the apparatus below to measure the half-life of a radioactive sample.

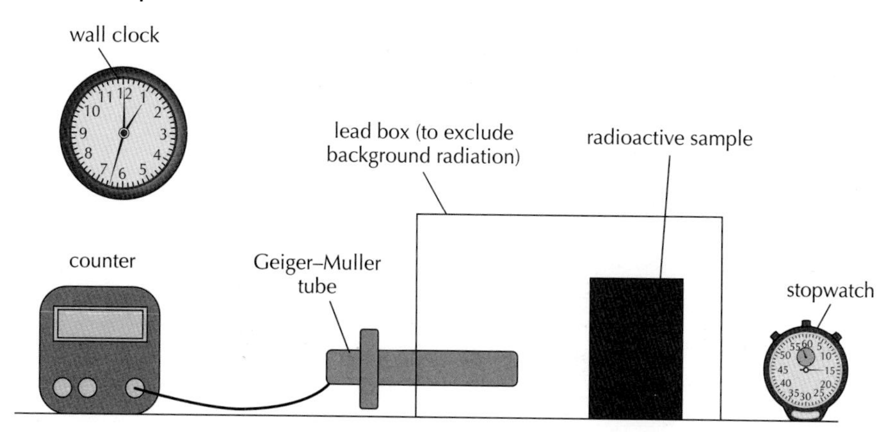

(a) (i) Describe how the technician could use the equipment to measure the half-life of the radioactive sample. Your description should include

- the apparatus required
- the measurements taken
- how the half-life is calculated. **3**

MARKS Do not write in this margin

(ii) A graph of count rate against time for the source is shown. The count rate has been corrected for background radiation.

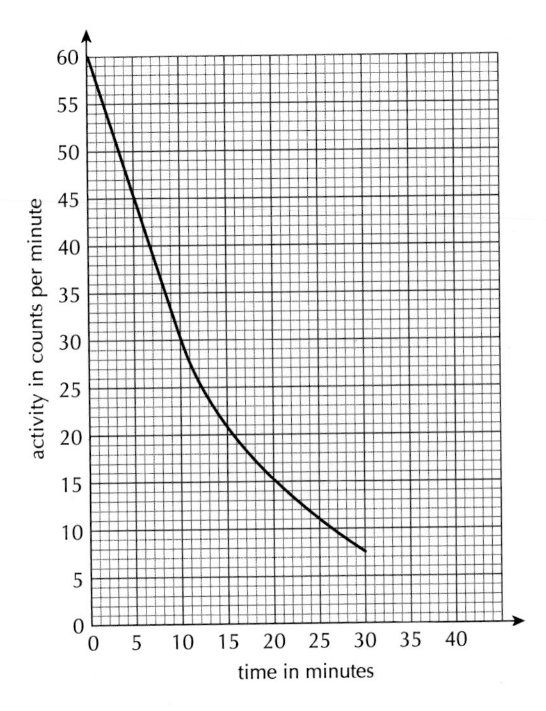

Use the graph to determine the half-life of the radioactive sample. **2**

Space for working and answer

(b) State **two** factors that can affect the background radiation level. **2**

(c) The radioactive sample emits alpha particles. What is an alpha particle? **1**

MARKS
Do not write in this margin

13. Nuclear power stations can be used to produce electricity for the National Grid.

In the nuclear reactor of a power station, a neutron strikes the nucleus of a uranium atom. The uranium nucleus splits into two smaller parts and three neutrons are released.

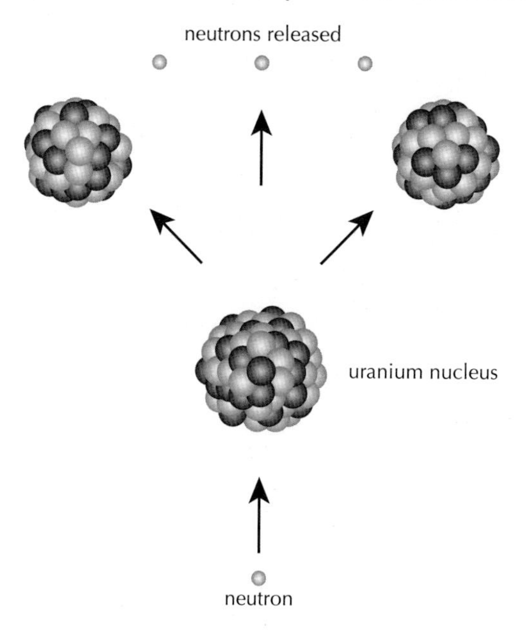

(a) State the name of this type of nuclear reaction. **1**

(b) A worker at the nuclear power station is responsible for removing the control rods from the nuclear reactor.

The worker's hand receives an average absorbed dose of 0·06 μGy each time a control rod is handled. The radiation weighting factor of the radiation is 20.

Calculate the equivalent dose received by the worker.

Space for working and answer **3**

(c) In nuclear reactions, mass is converted to energy. The energy, in Joules, released in a nuclear reaction can be calculated using the equation:

$$E = mc^2$$

where m is the mass converted to energy in kilograms and c is the speed of light in air in metres per second.

The total mass before the nuclear reaction is $387{\cdot}497 \times 10^{-27}$ kg. The total mass after the nuclear reaction is $386{\cdot}822 \times 10^{-27}$ kg.

Calculate the energy released in the reaction.

Space for working and answer **4**

[END OF QUESTION PAPER]

Practice paper B

Practice paper B

N5 Physics

Practice Papers for SQA Exams

Physics Section 1

Fill in these boxes:

Name of centre

Town

Forename(s)

Surname

Try to answer all of the questions in the time allowed.

Total marks — 135

Section 1 — 25 marks

Section 2 — 110 marks

Read all questions carefully before attempting.

You have 2 hours 30 minutes to complete this paper.

Write your answers in the spaces provided, including all of your working.

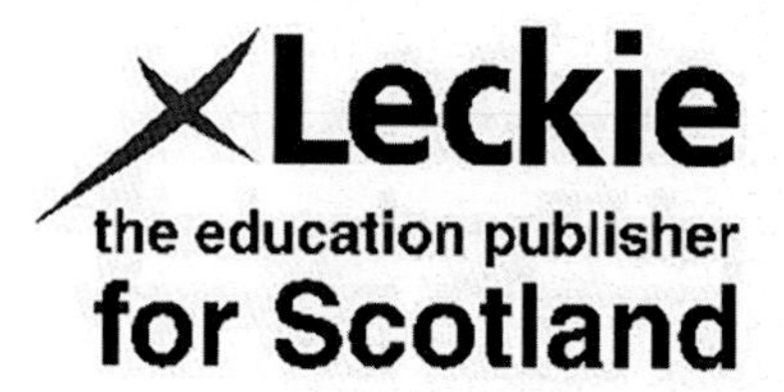

SECTION 1

Objective Test

1. Which row contains vector quantities only?

A	speed	distance	time
B	speed	displacement	acceleration
C	velocity	force	displacement
D	velocity	acceleration	displacement
E	velocity	distance	acceleration

2. A walker follows the route shown in the diagram.

Which row in the table shows the total distance travelled and the magnitude of the final displacement?

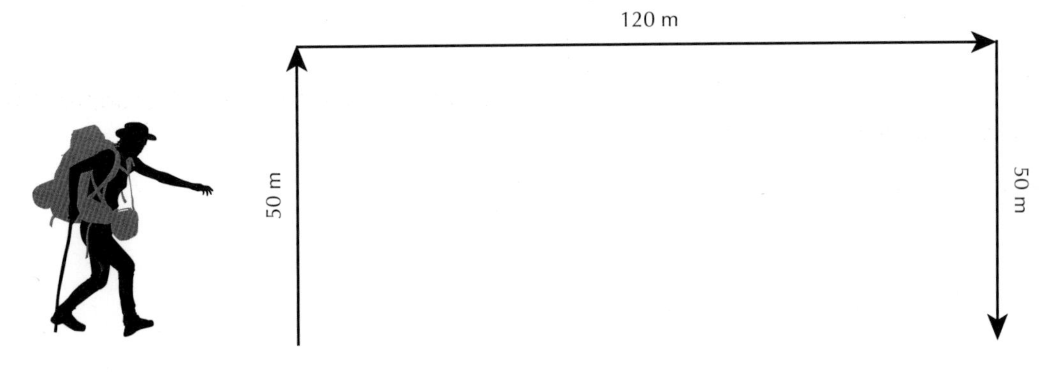

	Total distance **(m)**	***Final displacement*** **(m)**
A	120	50
B	120	220
C	170	120
D	220	120
E	220	220

3. A bus slows down as it approaches a set of traffic lights. The speed-time graph of the bus's motion is shown.

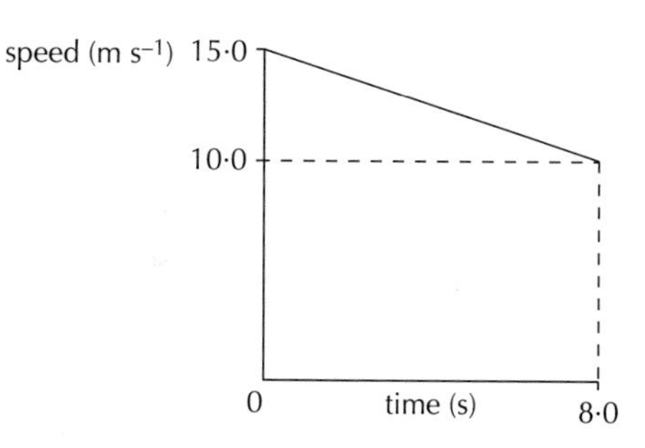

The distance travelled by the bus is

A 80 m

B 100 m

C 120 m

D 140 m

E 160 m.

4. A car is travelling at 12 m s^{-1} when the driver sees a red light ahead. The driver applies the brakes and the car slows to a stop in 6·6 s.

The deceleration of the car is

A - 1·80 m s^{-2}

B 0·55 m s^{-2}

C 1·80 m s^{-2}

D 5·40 m s^{-2}

E 79·2 m s^{-2}

5. A student writes the following statements about mass and weight.

I The mass of an object is different on Mars than it is on Earth.

II The mass of an object is the same on Mars as it is on Earth.

III The weight of an object is the same on Mars as it is on Earth.

Which of these statements is/are correct?

A I only

B II only

C III only

D I and III only

E II and III only

6. A toy car is travelling along a horizontal surface as shown.

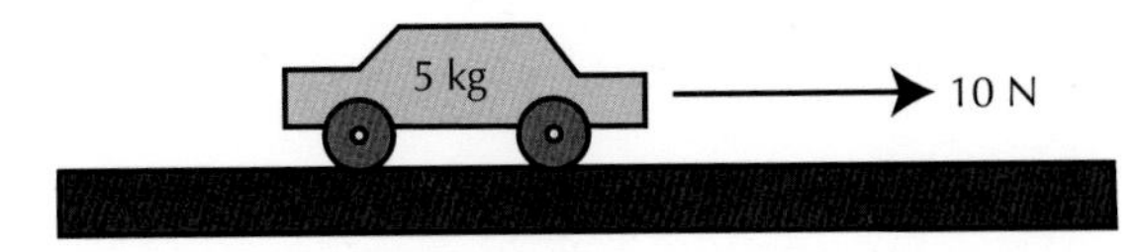

The mass of the car is 5 kg. The car is travelling at a constant speed.

The force of friction acting on the car is

A 0 N

B 2 N

C 5 N

D 10 N

E 50 N.

7. The period of a geostationary satellite is

A 1 hour

B 1 day

C 1 week

D 1 month

E 1 year

8. Electric current is the flow of

A Protons

B Neutrons

C Electrons

D Electrons and neutrons

E Protons and neutrons.

9. A range of potential differences are applied across a resistor and the corresponding currents are measured.

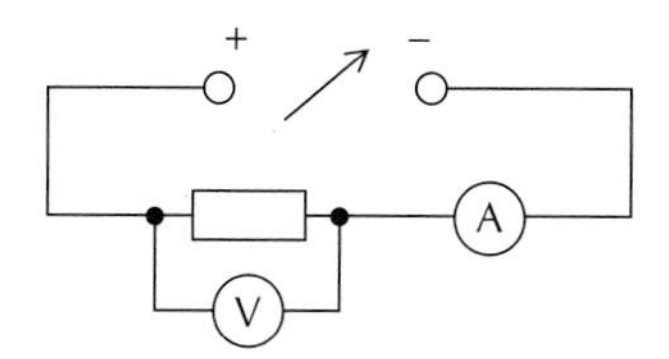

Assuming the temperature of the resistor remains constant, which of the following shows the current-voltage graph obtained?

A
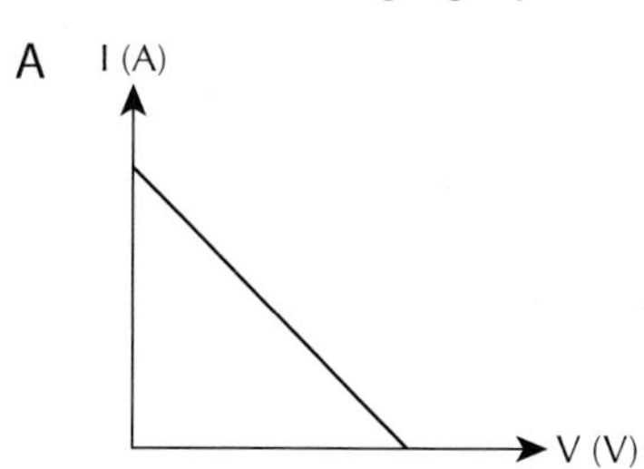

B
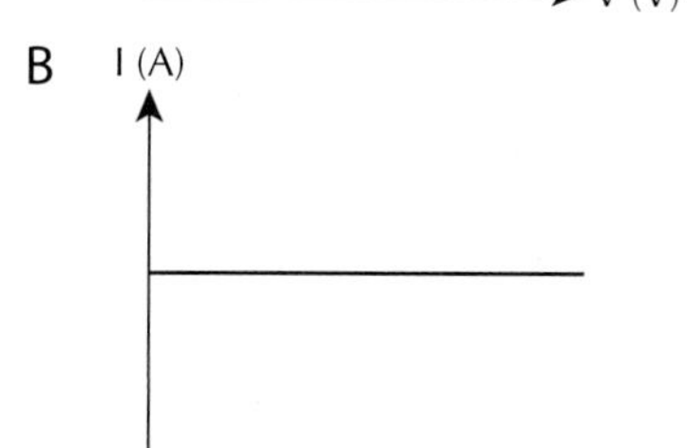

C
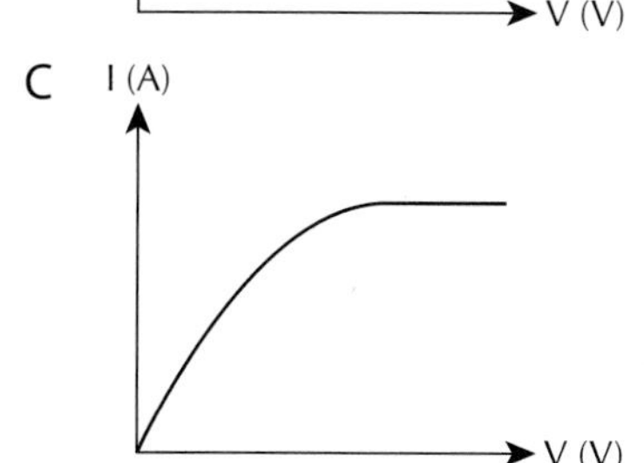

D
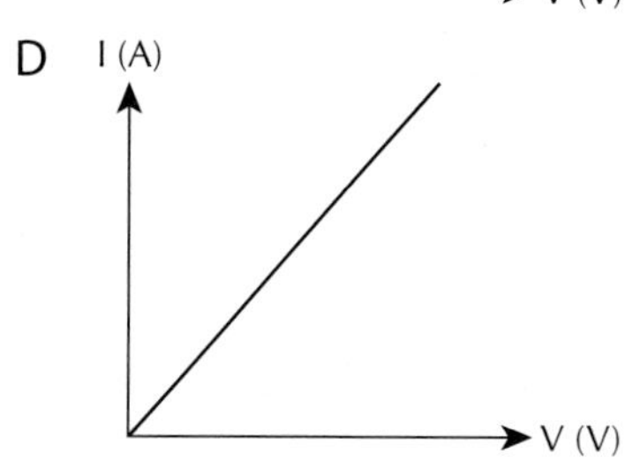

E
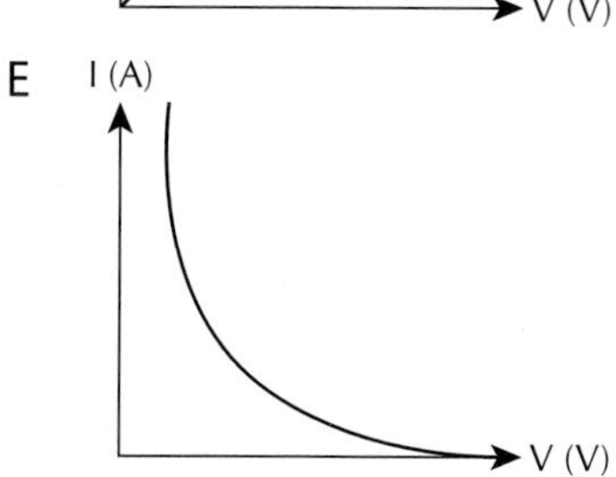

10. Consider the following circuit.

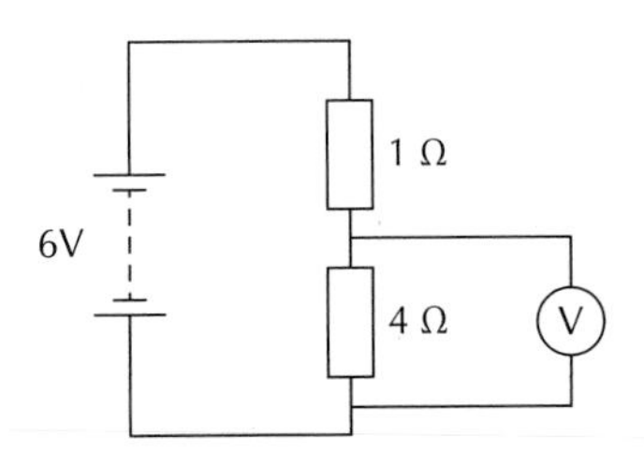

The reading on the voltmeter is

A 1·0 V

B 1·2 V

C 4·8 V

D 5·0 V

E 6·0 V.

11. Two identical resistors are connected with four ammeters as shown.

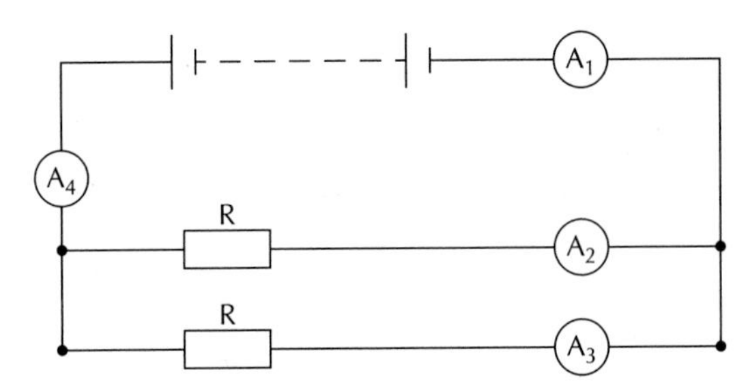

The reading on A_2 is 0·2 A.

Which row shows the readings on A_1, A_3 and A_4?

	Ammeter A_1	***Ammeter A_3***	***Ammeter A_4***
A	0·2 A	0·2 A	0·2 A
B	0·2 A	0·4 A	0·6 A
C	0·4 A	0·4 A	0·2 A
D	0·4 A	0·2 A	0·6 A
E	0·4 A	0·2 A	0·4 A

12. Which of the following equations can be used to find the power dissipated by a resistor?

I $P = IV$

II $P = I^2 R$

III $P = \frac{V^2}{R}$

A I only

B II only

C III only

D II and III only

E I, II and III

13. A heater is immersed in a substance. The heater is then switched on.

The graph shows the temperature of the substance over a period of time.

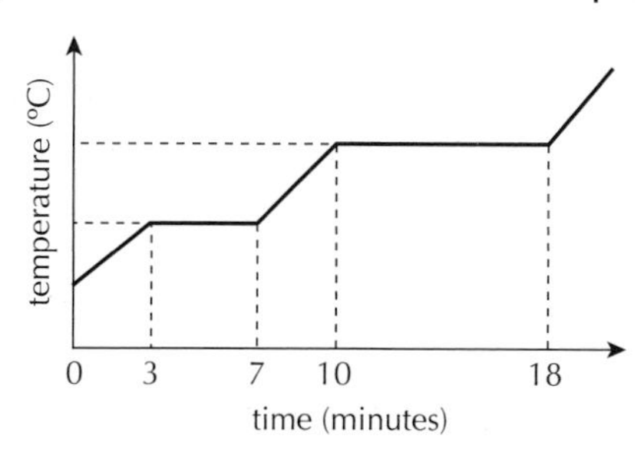

After 10 minutes the substance is

A in the solid state

B in the liquid state

C a mixture of solid and liquid

D in the gaseous state

E a mixture of liquid and gas

14. An aircraft is travelling at a constant height above the Earth. The air pressure at this height is $0{\cdot}2 \times 10^5$ Pa. The inside of the aircraft is maintained at a pressure of $1{\cdot}0 \times 10^5$ Pa.

The area of an external door is 2 m^2. What is the outward force produced on the door?

A $0{\cdot}4 \times 10^5$ N

B $0{\cdot}5 \times 10^5$ N

C $1{\cdot}6 \times 10^5$ N

D $2{\cdot}0 \times 10^5$ N

E $2{\cdot}4 \times 10^5$ N

15. The pressure of a fixed mass of gas is 200 kPa at a temperature of −17 °C. The volume of the gas remains constant.

At what temperature would the pressure of the gas be 300 kPa?

A 111 °C

B 162 °C

C 299 °C

D 384 °C

E 435 °C

16. The end of a syringe is sealed. The air inside is now trapped.

The plunger is pushed in slowly. The pressure of the air increases.

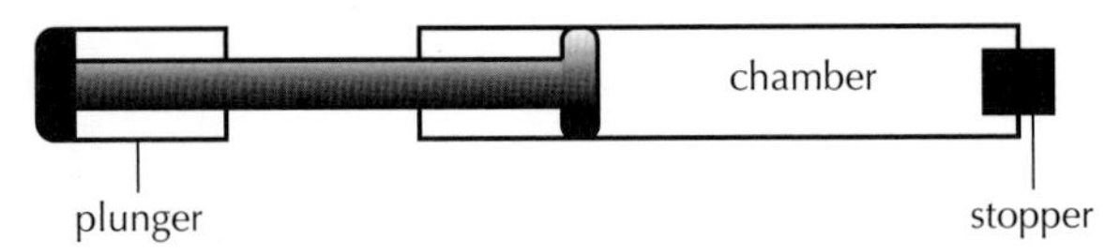

Which of the following explain(s) why the pressure increases, assuming the temperature remains constant?

I The air particles collide more often with the walls of the syringe.

II The air particles increase their average speed.

III The air particles strike the walls of the syringe with greater force.

A I only

B II only

C III only

D I and II only

E I and III only

17. Which of the following is a longitudinal wave?

A Water wave

B Light wave

C Sound wave

D Gamma ray

E Radio wave

18. A water wave is shown below.

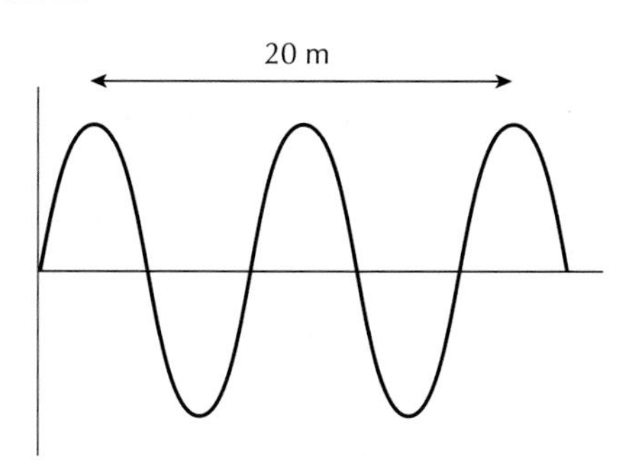

The speed of the wave is $2{\cdot}5\ \text{m s}^{-1}$.

The frequency of the wave is

A 0·125 Hz

B 0·25 Hz

C 4·0 Hz

D 8·0 Hz

E 25 Hz.

19. Which of the following electromagnetic waves has a lower frequency than infrared and a shorter wavelength than radio?

A X-rays

B Gamma rays

C Visible light

D Ultraviolet

E Microwave

20. A ray of light is incident on a glass block as shown.

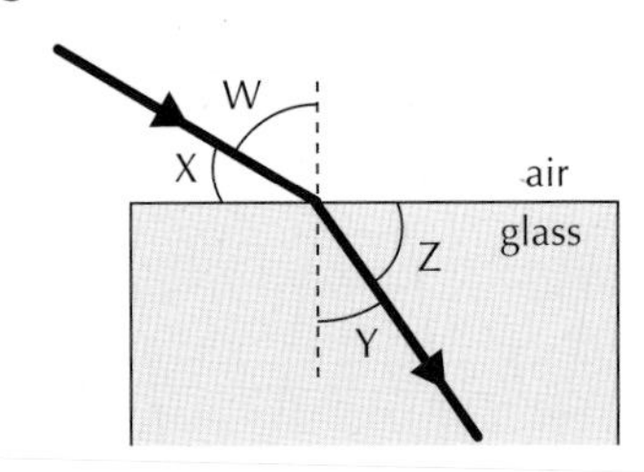

Which row correctly identifies the angle of incidence and the angle of refraction?

	Angle of incidence	*Angle of refraction*
A	W	Y
B	W	Z
C	X	Y
D	X	Z
E	Y	W

21. The diagram below shows a simple model of the atom.

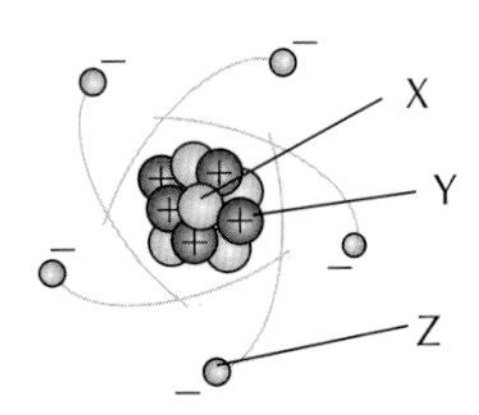

Which row in the table identifies particles X, Y and Z?

	X	*Y*	*Z*
A	neutron	proton	electron
B	proton	neutron	electron
C	neutron	electron	proton
D	proton	electron	neutron
E	electron	proton	neutron

22. A patient's thyroid gland has a mass of 0·04 kg.

The gland is exposed to radiation and absorbs 0·2 μJ of energy in 3 minutes.

The absorbed dose is

A 0·005 Gy

B 0·12 Gy

C 5 μGy

D 5 Gy

E 15 MGy.

23. The following statements are made about nuclear fission.

I The nucleus splits into smaller parts.

II Energy is released.

III Nuclei join together.

Which of the statements is/are correct?

A I only

B II only

C III only

D I and II only

E II and III only

24. The average annual background radiation in the UK in millisieverts per year is

A 0·022

B 0·22

C 2·2

D 22

E 220

25. Which of the following describes the term ionisation?

A An atom losing a proton

B A nucleus emitting a gamma ray

C An atom losing an electron

D A nucleus emitting an alpha particle

E An atom losing a neutron

N5 Physics

Practice Papers for SQA Exams | Physics Section 2

Fill in these boxes:

Name of centre

Town

Forename(s)

Surname

Try to answer all of the questions in the time allowed.

Total marks — 135

Section 1 — 25 marks

Section 2 — 110 marks

Read all questions carefully before attempting.

You have 2 hours 30 minutes to complete this paper.

Write your answers in the spaces provided, including all of your working.

MARKS Do not write in this margin

SECTION 2

1. A spacecraft is on the surface of the Moon.

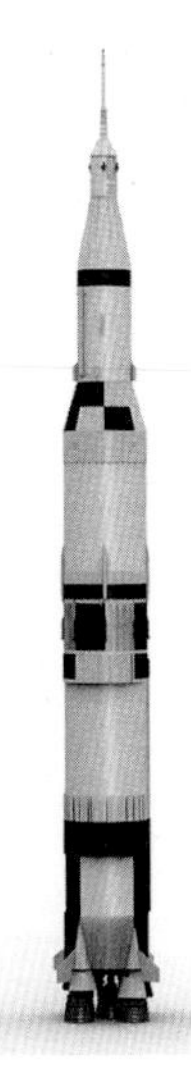

The spacecraft has a total mass of $2 \cdot 8 \times 10^6$ kg.

The spacecraft's engines produce a total force of $3 \cdot 5 \times 10^7$ N.

(a) (i) Calculate the weight of the spacecraft on the surface of the Moon. **3**

Space for working and answer

(ii) Sketch a diagram showing the forces acting on the spacecraft immediately after take-off from the Moon. You must name the forces and show their direction. **2**

(iii) Calculate the acceleration of the spacecraft as it takes off from the Moon. **4**

Space for working and answer

MARKS Do not write in this margin

(b) An identical spacecraft is launched from the surface of the Earth. The mass of the spacecraft and the engine force are the same as before.

Is the acceleration of the spacecraft as it takes off from the Earth greater than, less than, or equal to the acceleration as it takes off from the Moon?
You must justify your answer by calculation.

Space for working and answer **4**

MARKS
Do not write in this margin

2. A skateboarder is practising on a ramp. The mass of the skateboarder and the board is 70 kg.

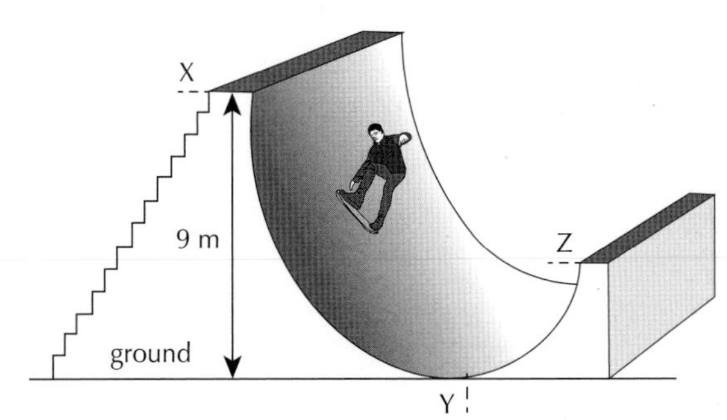

(a) Calculate the increase in potential energy of skateboarder and board in moving from the ground to position X. **3**

Space for working and answer

(b) The skateboarder moves from X to Z.

(i) At what point on the ramp is the kinetic energy of the skateboarder greatest? **1**

(ii) The vertical speed of the skateboarder at Z is 5·5 ms^{-1}.
Calculate the maximum height above Z the skateboarder can rise to. **5**

MARKS
Do not write in this margin

3. A volleyball player strikes a ball. The ball leaves the hand horizontally at 12 m s^{-1}. It hits the ground at a distance of 5·8 m from the point where it was struck.

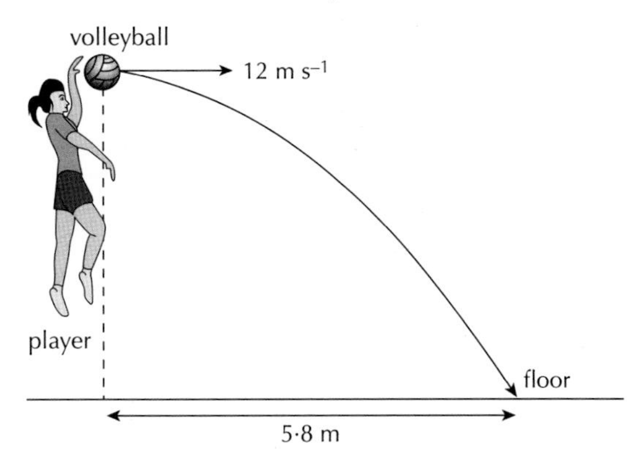

Assume that air resistance is negligible.

(a) Explain why the shape of the path taken by the ball is curved. **2**

(b) Calculate the time of flight of the ball.

Space for working and answer **3**

(c) Calculate the vertical speed of the ball as it reaches the ground.

Space for working and answer **3**

(d) Sketch a graph of vertical speed against time for the ball. Numerical values are required on both axes. **2**

MARKS
Do not write in this margin

4. The 2018 World Cup will be held in Russia. The matches will be transmitted to the UK by microwaves. The microwaves are transmitted via a geostationary satellite positioned 36 000 km above the Earth.

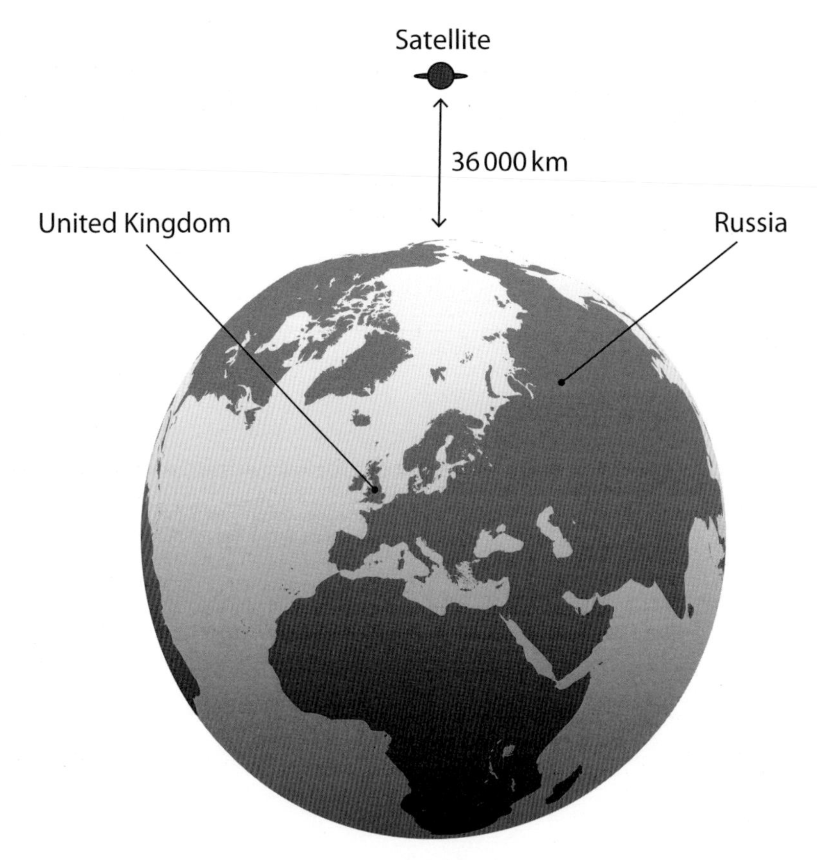

(a) What is the period of a geostationary satellite? **1**

(b) The frequency of the microwaves used in the transmission is 220 GHz.

Calculate the wavelength of the microwaves used in the transmission. **3**

Space for working and answer

MARKS
Do not write in this margin

5.

Fraunhofer lines

The emission spectrum of the Sun has a number of dark lines in it. The dark lines are caused by absorption by cooler gases just above the hot visible surface that we see. The lines are called Fraunhofer lines after Josef von Fraunhofer (Bavaria, 1787–1826), who discovered the dark lines. He developed a device called a spectroscope to view the absorption lines. The absorption lines indicate the elements that are present at the Sun's surface.

The intensity of the absorption lines for an element can tell us how much of the element is present: the more of the element that is at the surface, the more absorption takes place and the darker the line. Such measurements show that the Sun's atmosphere consists of 72% hydrogen, 26% helium and 2% heavier elements.

(a) Name the device used by Fraunhofer in the discovery of absorption lines. **1**

(b) Explain how the absorption lines are used to determine how much of each element is present. **1**

(c) A line spectrum from the Sun is shown below along with the line spectra of the elements sodium, helium, hydrogen and nitrogen.

Line spectrum from Sun

Sodium

Helium

Hydrogen

Nitrogen

Identify the elements present in the Sun. **2**

MARKS Do not write in this margin

6. An LDR is used as a light sensor in the circuit below. When the light level falls below a certain value, the output causes a switching circuit to turn on a street light.

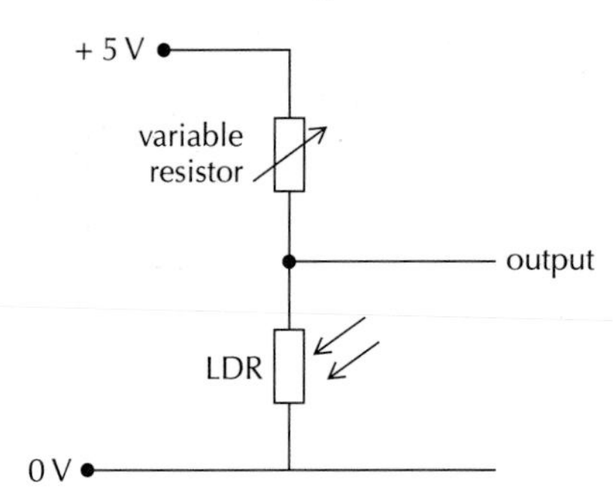

(a) When the voltage across the LDR is 0·7 V, the circuit causes the street light to switch on.

The resistance of the variable resistor is set at 8600 Ω.

Calculate the resistance of the LDR when the voltage across the LDR is 0·7 V.

Space for working and answer **4**

(b) The graph shows how the resistance of the LDR changes with temperature.

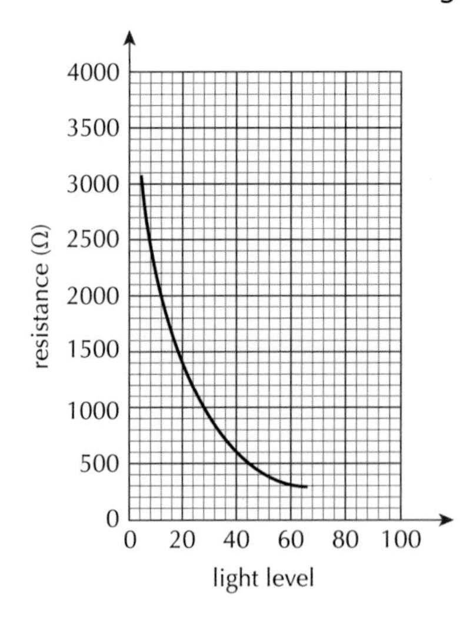

(i) State what happens to the resistance of the LDR as the light level increases. **1**

(ii) Use the graph to determine the light level at which the street light is switched on. **1**

(c) The switching circuit is connected as shown. When there is a current in the relay coil, the relay switch closes.

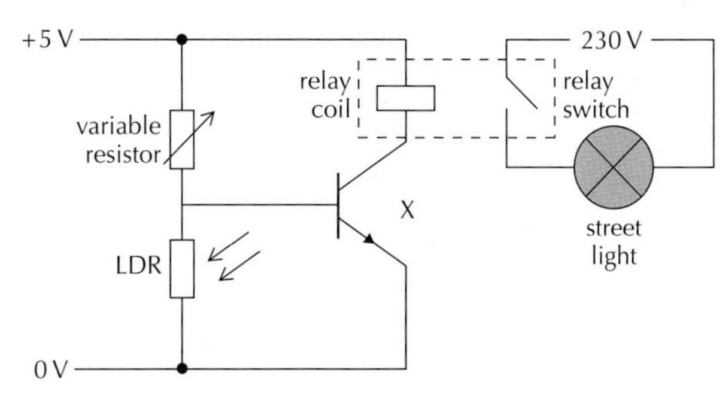

(i) Name component X. **1**

(ii) Explain how the circuit operates to switch on the street light. **3**

7. An electric kettle is filled with water.

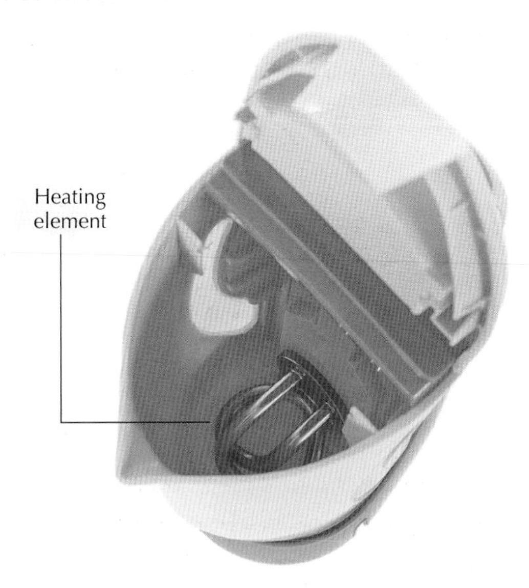

The heating element in the kettle is used to raise the temperature of the water. The heating element is rated at 240 V, 2 kW.

(a) The heating element is switched on for 40 s.

Calculate the electrical energy supplied to the heating element.

Space for working and answer **3**

(b) The mass of water inside the kettle is 0·3 kg. The temperature of the water rises from 25 °C to 84 °C during the time the heating element is switched on.

Calculate the heat energy gained by the water.

Space for working and answer **4**

(c) Explain why the heat energy gained by the water is less than the electrical energy supplied to the heating element. **2**

MARKS
Do not write in this margin

8. A student is investigating the motion of a water rocket. The water rocket is made from a plastic bottle containing some water. When air is pumped into the bottle the pressure becomes large enough that the bottle is launched upwards.

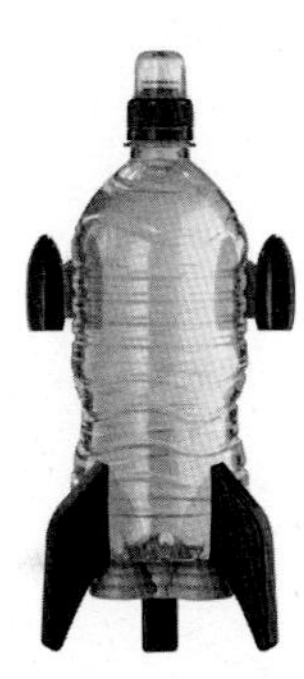

(a) Before launch, the water rocket rests on three fins on the ground. The area of each fin in contact with the ground is 8×10^{-4} m. The mass of the rocket before launch is 0·77 kg.

Calculate the total pressure exerted on the ground by the fins. **6**

Space for working and answer

(b) Explain fully why the rocket rises as water is forced out of the bottle **2**

MARKS Do not write in this margin

9. A pulsar is a rapidly spinning star that emits radio waves.

The pulsar PSR B1257 + 12 is $9{\cdot}46 \times 10^{18}$ m from the Earth.

The radio waves are received by a telescope on Earth.

(a) What is the speed of radio waves? **1**

(b) Calculate the time taken for the radio waves to reach Earth.

Space for working and answer **3**

(c) The period of rotation of PSR B1257 + 12 is 6·2 ms.
Calculate the number of rotations in 1 second.

Space for working and answer **3**

(d) As they grow older, all pulsars slow down.

Astronomers measuring another pulsar in the Crab Nebula have created an equation to predict the spin rate of the pulsar in the future.

$P = 0{\cdot}033 + 0{\cdot}000013T$

where P is the period of rotation of the Crab Nebula pulsar in seconds and T is the number of years since today.

Calculate the period of rotation of the Crab Nebula Pulsar 10 000 years in the future.

Space for working and answer **2**

MARKS Do not write in this margin

10. A news report states:

'We're probably being too optimistic about finding life elsewhere in the Universe – it's very unlikely.'

Using your knowledge of physics, comment on the above statement. **3**

11. Infrared radiation is used to send information via fibre optic cables.

(a) The diagram shows an infrared ray incident on a glass fibre.

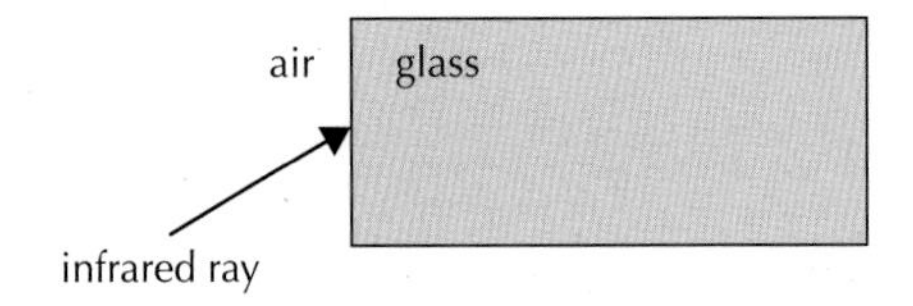

Complete the diagram to show the path of the infrared ray as it enters the glass.

Indicate on your diagram the normal, the angle of incidence and the angle of refraction. **2**

(b) Infrared radiation is part of the electromagnetic spectrum.

A section of the electromagnetic spectrum is shown in the diagram below.

Infrared radiation	Visible light	Ultraviolet light	**A**	Gamma radiation

(i) Identify radiation A. **1**

(ii) Which of these electromagnetic waves has the greatest energy? **1**

(iii) State **one** property that all electromagnetic waves have in common. **1**

MARKS Do not write in this margin

12. (a) Rutherfordine is a mineral that contains Uranium.

The activity of 1·0 kg of pure Rutherfordine is $1{\cdot}3 \times 10^8$ decays per second.

A sample of mass 0·4 kg contains 50% Rutherfordine. The other 50% is not radioactive.

Calculate the activity of the sample in becquerels.

Space for working and answer **4**

(b) The equivalent dose limit for a worker is 0·2 mSv per day.

The worker is in contact with the radiations for 8 hours each day.

The table below gives information on the radiation exposure.

Type of radiation	***Absorbed dose per hour***
Gamma	0·1 μSv
Fast neutrons	1 μSv
Slow neutrons	4 μSv

Show, by calculation, whether the equivalent dose limit is exceeded.

Space for working and answer **5**

MARKS
Do not write in this margin

13. Workers in a factory assembling smoke detectors experience a higher equivalent dose than that due to background radiation.

(a) State two sources of background radiation **2**

(b) The workers assembling the smoke detectors should not receive an equivalent dose greater than 15 mSv per year above background level.

(i) State the annual effective dose limit for a radiation worker **1**

(ii) A worker in the factory assembles 10 000 smoke detectors in a year. An absorbed dose of 6×10^{-8} Gy is received by the worker in assembling one detector. The radiation weighting factor of the radiation is 20.

Show by calculation if the worker will receive an equivalent dose that is greater than 15 mSv per year. **5**

Space for working and answer

MARKS
Do not write in this margin

14. The photoelectric effect was discovered towards the end of the 19th century. Experiments showed that electrons could be emitted from the surface of a metal by illuminating the metal with light.

The electrons are only emitted from the surface of the metal if the light is above a certain frequency. This is known as the threshold frequency.

The energy needed to remove an electron from the surface of a metal is called the work function. The work function, in Joules, of a metal is given by the equation:

$$E = hf_0$$

where h is Planck's constant ($h = 6{\cdot}63 \times 10^{-34}$ J s) and f_0 is the threshold frequency.

(a) The table below gives the work function of different metals.

Metal	***Work function* (J)**
Zinc	$6{\cdot}9 \times 10^{-19}$
Gold	$7{\cdot}8 \times 10^{-19}$
Sodium	$3{\cdot}6 \times 10^{-19}$
Potassium	$3{\cdot}2 \times 10^{-19}$

An unknown metal, metal X, is found to have a threshold frequency of $1{\cdot}18 \times 10^{15}$ Hz. Identify metal X.

Space for working and answer **3**

(b) Calculate the longest wavelength of light that would allow an electron to be emitted from the metal potassium. **3**

MARKS
Do not write in this margin

15. When white light is passed through a glass prism, a spectrum of colours is produced.

A student makes the statement:

'The colours are produced by the glass prism.'

Use your knowledge of physics to explain why the student is incorrect. **3**

[END OF QUESTION PAPER]